This book is protected by copyright

Legal information

© 2023
Author and Editor: M.Eng. Johannes Wild
A94689H39927F
Email: 3dtech@gmx.de

The complete imprint of the book can be found on the last pages!

This work is protected by copyright

The work, including its parts, is protected by copyright. Any use outside the narrow limits of copyright law without the consent of the author is prohibited. This applies in particular to electronic or other reproduction, translation, distribution and making publicly available. No part of the work may be reproduced, processed or distributed without written permission of the author! All rights reserved.

All information contained in this book has been compiled to the best of our knowledge and has been carefully checked. However, the publisher and the author do not guarantee the timeliness, accuracy, completeness and quality of the information provided. This book is for educational purposes only and does not constitute a recommendation for action. The use of this book and the implementation of the information contained therein is expressly at your own risk. In particular, no warranty or liability is given for damages of a material or immaterial nature on the part of the author and publisher for the use or non-use of information in this book. This book does not claim to be complete or error-free. Legal claims and claims for damages are excluded. The operators of the respective Internet sites referred to in this book are exclusively responsible for the content of their site. The publisher and the author have no influence on the design and contents of third party internet web sites. The publisher and author therefore distance themselves from all external content. At the time of use, no illegal content was present on the websites. The trademarks and common names cited in this book remain the sole property of the respective author or rights holder.

Thank you so much for choosing this book!

Foreword

Thank you very much for choosing this book!

If you are looking for a practical guide for the awesome and multifunctional Arduino mini-PC, then you have come to the right place and are well advised with this book: "Arduino | Step by Step"! I am an engineer (M.Eng.) and would like to introduce you to the world of Arduino in a very simple way. You will learn both the theoretical basics for handling an Arduino, and the practical application through awesome and exciting example DIY projects. This book offers you an easy-to-understand, intuitively structured and practical introduction to the world of the mini-PC! We will work exclusively with the Arduino Uno in this book, as it is ideally suited for beginners.

From the basics of electrical engineering, setting up components for the Arduino board and building circuits, getting to know the software, up to programming and creating some projects (such as creating an SOS signal, creating temperature or light dependent Arduino systems, or password protected and remote-controlled Arduino systems). All of this is included in this book, giving you all the knowledge you need to get started.

This basic book is aimed specifically at all those who have no or only very primitive prior knowledge of Arduino. No matter what age you are, what profession you have, whether you are a pupil, student, or pensioner. This book is for everyone who wants to get familiar with the fascinating topics: Electronics, Arduino and programming.

The goal of this book is to teach you what an Arduino is, how it works, and how to use it for great projects. It is a book that provides an understanding of electrical engineering basics as well as the basics of programming and building circuits for an Arduino.

So, in this Arduino basic course you will learn everything you need to know as a beginner about the world of the mini-PC, its programming and circuit design! Best to take a look at the book and get your copy!

Table of contents

Foreword 2

Table of contents 3

1 Introduction: scope of the book 5

2 First steps with Arduino 6

3 Background knowledge - Fundamentals of electrical engineering 9

    3.1 Introduction to electricity and digital electronics 9

        3.1.1 Electricity 9

        3.1.2 Circuit 10

        3.1.3 Digital electronics 11

    3.2 Important components in electronics 12

        3.2.1 The diode and the light-emitting diode (LED) 12

        3.2.2 The transistor 13

        3.3.3 The capacitor 14

        3.3.4 The resistance 15

    3.3 The multimeter: Current and voltage measurement in practice 16

    3.4 Embedded Systems 18

4 The Arduino Board Hardware 21

5 The Arduino Software (IDE) 25

6 Writing a program code for the Arduino 26

    6.1 Introduction to programming 26

        6.1.1 Block-based programming 26

        6.1.2 Basics for using the Arduino IDE 29

        6.1.3 Libraries 30

        6.1.4 Serial monitor 35

    6.2 Basics for text-based Arduino programming 35

Arduino | Step by step

    6.2.1 Structure of an Arduino code 36

    6.2.2 Syntax (program structure) for Arduino code 37

    6.2.3 Basic operators in Arduino programming 37

    6.2.4 Basic data types for Arduino programming 38

    6.2.5 Constants and variables 39

    6.2.5 Controlling the operation of the Arduino 41

    6.2.6 Functions 42

  6.3 Connect to the Arduino Board and Upload Code (Sketch) 46

7 Arduino DIY Projects 48

  7.1 Project 1: A flashing LED and an SOS signal 48

  7.2 Project 2: Temperature-based LED light 53

  7.3 Project 3: Light-dependent control of a motor (blind motor) 59

  7.4 Project 4: Gas detection alarm 63

  7.5 Project 5: Password protected mechanical system 68

  7.6 Project 6: Remote unlocking mechanism 81

Closing words 88

Imprint of the author / publisher 92

1 Introduction: scope of the book

What to expect and what you will learn in this book

In this Arduino Beginner's Guide, you will find an introduction to the use of the mini-PC Arduino in theory and practice. I, as an engineer, will share with you my knowledge from study and practice step by step in this book. This will help you to achieve an optimal learning success with theoretical basics on the one hand, but on the other hand especially with practical examples. In the first chapters you can expect basic background knowledge of electrical engineering and electronics, then we will come to theoretical topics regarding Arduino and will finally practice our learnings in the form of exciting projects.

In this course, aimed specifically at beginners, you will learn all the basics you need to know when working with an Arduino. We will work exclusively with the Arduino Uno in this book, as it is ideally suited for beginners.

<u>In brief, this course will teach you the following in detail:</u>

- The basic terms and components of electrical engineering as background knowledge

- The structure of an Arduino Uno board and how to use it

- What is the Arduino IDE, what is it used for and how is it structured?

- Programming basics: block-based programming

- Programming basics: text-based programming

- How to create a system with an Arduino and how to write the required program code

- Hands-on learning based on exciting DIY projects: SOS signal with LED, temperature-based LED control, light-dependent control of a motor, gas detection alarm, password-protected system, remote-controlled system.

- and much more!

Be excited! Let's go!

2 First steps with Arduino

Arduino is an electronic platform consisting of hardware and software that is very user-friendly and was created as part of an open-source project. The term open source is generally characterized by the fact that the software is freely available, active participation of users is desired and there are no restrictions on use. Simply put, an Arduino is nothing more than a small and very simple PC or microcontroller that is capable of taking input signals, processing them internally, and then converting them into corresponding output signals. An input signal could be e.g., sunlight falling on a sensor. The corresponding output signal could, for example, control a motor. This mini-PC can be purchased in the very puristic appearance of a circuit board, equipped with electronic components, either individually or as a set.

There are different Arduino boards, modules, and beginner sets.

The following Arduino boards are recommended to get started:

- Arduino Uno
- Arduino Nano
- Arduino Leonardo
- Arduino Micro

A good overview of all products and a way to order them can be found on the official website: https://www.arduino.cc/en/main/products . You can either order the Arduino products we need in this course via this official site or buy them via Amazon or eBay. By the way, an Arduino Uno board is available from about $20, a complete starter set from about $70.

In this book, we will mainly deal with the Arduino Uno board and use it for the projects. We will also need other components such as LEDs, resistors, sensors (e.g., infrared sensor), actuators (e.g., motor) for the projects. Which components in particular are needed, you will find in the further course and in each project clearly listed. For this, it is recommended to buy the so-called "Arduino Starter Kit for Beginners" (available on Amazon, eBay...), or any other startet-set, or in addition to the Arduino Uno board a sensor / module set, which contains the required components.

In order for the Mini-PC to know what to do with the previously mentioned input signals and what output signals we would like to have, the Arduino board needs instructions. These instructions for the microcontroller are given by the user, i.e., by us, thanks to a program code. A programming language is used for this purpose. For programming and transmission, a special software is used, the Arduino Software (IDE). This can be downloaded online free of charge, e.g., here: https://www.arduino.cc/en/software.

Countless projects have already been realized with the Arduino microcontroller and this mini-PC is suitable for hobby projects, for prototyping or even for scientific projects. The Arduino community is spread all over the world. It includes students, engineers, hobbyists, artists, programmers, and so on. Millions of users have contributed to this open-source platform, and thanks to these contributions, a lot of knowledge has been accumulated to help professional and new users with their various projects. Arduino is specifically designed for users who need a simple and inexpensive platform for electronics and programming projects. Since Arduino is an open-source project, users can change anything they want or customize any function according to their needs.

Why should you choose an Arduino?

First of all, Arduino is very user-friendly, as it has a wide range of applications, even especially for beginners. Nevertheless, this platform is equally well suited for experienced users. Another reason is the large Arduino community that supports this open-source project. Arduino also works with both a Mac, Windows, or Linux operating system. Thanks to the simplicity of the platform and the wide range of functions, an Arduino can be used by both a child and a retiree. Especially for beginners, Arduino is optimal for creating projects from simple to complex due to its ease of use and the large amount of data that already exists.

There are also other microcontroller platforms and competing products. Some of them that are similar to Arduino are called for example: "Basic Stamp" by "Parallax", "BX-24", "Phi" and "Handy Board" and there are many other boards with similar functions. However, these microcontrollers use quite old-fashioned programming methods, the community is not as large as for Arduino and the instructions are not so easy for newcomers.

The following is a brief list of why Arduino is so great and why you are right to choose Arduino and this course:

Reasonable price

Arduino has a fair to reasonable price, and this is one of the main reasons for its worldwide success. The Arduino Uno board, for example, is already available for around $20. A starter set from about $70.

Cross-platform

As with many major platforms, most microcontrollers only work with Windows. They lack support for systems like Mac and Linux. Arduino, on the other hand, runs with all systems.

Easy to program

Probably the most important point. An Arduino is effortless to use and program. The software for it (Arduino IDE) is very user-friendly. This helps especially beginners, but also young people or pensioners, to get familiar with the program very easily and playfully. Nevertheless, Arduino also offers the possibility to perform complex projects and programming, so it is also a great platform for advanced users.

Open-Source Software

Everyone can contribute to this great project. Every user can create new libraries (we will learn what that is) and make them available to other users as well.

Open-Source Hardware

The Arduino hardware is also open source and can be modified by any user. Through a kind of plug-and-play system and through a so-called "breadboard", modules can be added, and a variety of different projects can be implemented. It is a kind of modular system.

3 Background knowledge - Fundamentals of electrical engineering

3.1 Introduction to electricity and digital electronics

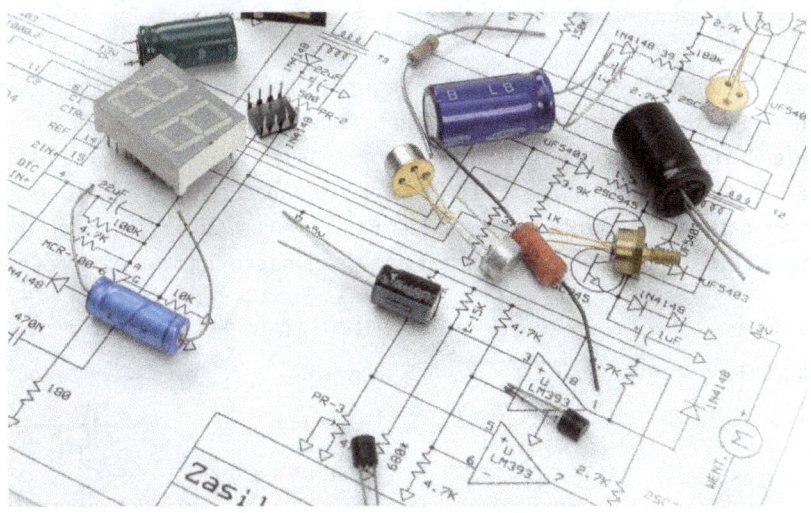

3.1.1 Electricity

Electricity is created by electrons flowing from a place with higher potential (higher energy) to a place with lower potential (lower energy). It can be relatively well imagined by means of a waterfall. The water (represents the electrons) flows from the top point of the waterfall (high potential, high potential energy) to the bottom point of the waterfall (low potential, lower potential energy). The potential energy is transformed into kinetic energy during this process, that's why it "loses" this high-energy state in the process (but actually this energy is transformed, as said before). Similarly, the electron wants to flow from a place with higher voltage (high potential) to a place with lower voltage (low potential).

Voltage is the unit of electrical energy "generated" by the battery. The battery or any other voltage source has two terminals. One terminal is called the negative terminal and the other terminal is called the positive terminal. At the positive terminal, the voltage potential is higher than compared to the negative side. Thus, the current flows from the positive side (plus pole) to the negative side (minus pole), considering the technical direction of current.

You can think of a battery or other power-generating source as functioning as a pump. A battery, for example, "generates" voltage or energy through an electrochemical reaction inside. This voltage or energy flows out of the positive pole in the form of electrons (these electrons symbolize the water molecules that are pumped out). To compensate for the "lost" electrons, the battery (similar to a suction pump) draws the same number of electrons back in through the negative pole.

3.1.2 Circuit

What is a circuit? Simply put, a circuit is an arrangement of different components with an electrically conductive connection between them. For an electrical circuit or circuit to work, you need an energy source / current source, such as a battery and a consumer, such as a light bulb, as well as connections between these two components, which are called conductors. In electrical engineering, these components are represented in a circuit or a circuit as symbol symbols as follows:

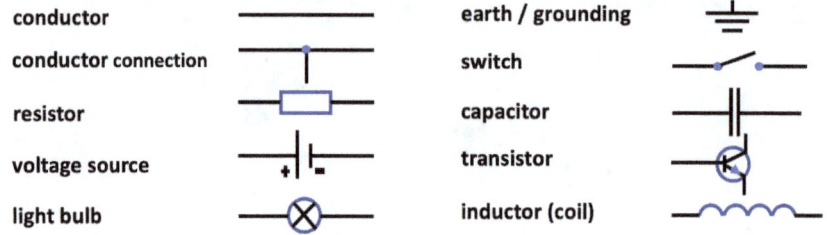

For a lamp, for example, to light up as shown in the following figure, the circuit must be closed, i.e., there must be a connection between the two poles (+ and -) of a power source (e.g., battery) and the incandescent lamp. If this is the case, current flows from one pole of the power source (e.g., battery) through the incandescent lamp and back to the other pole of the power source. When this connection is severed, e.g., by a switch, current no longer flows and the lamp no longer lights. In this case, it is called an open circuit. A short circuit occurs if the current can flow from one pole of the current source to the other pole unhindered and without first passing through an electrical component (e.g., through an uninsulated spot of a cable on a metal surface). This is because the current always takes the path of the least resistance.

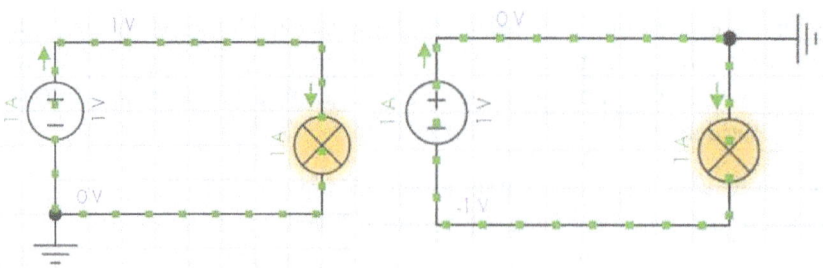

A circuit diagram is the basic concept of a circuit that can be drawn, for example, on a piece of paper or with the help of a computer program (see picture above).

Such a circuit diagram can also be made a bit more descriptive (see picture below). Circuit diagrams for the Arduino can be created best with the software from https://fritzing.org/download/ , which can be downloaded at the given URL for little money. At https://fritzing.org/learning/ you can also find many references and instructions for using the software. As we can see on the picture below, in this project for example a relay and a module are connected to an Arduino Uno via colored cables. The colors of the wires each have a meaning that helps to make a correct wiring. In this illustration, all red wires stand for the 5V signal and all black wires for the ground signal (0V).

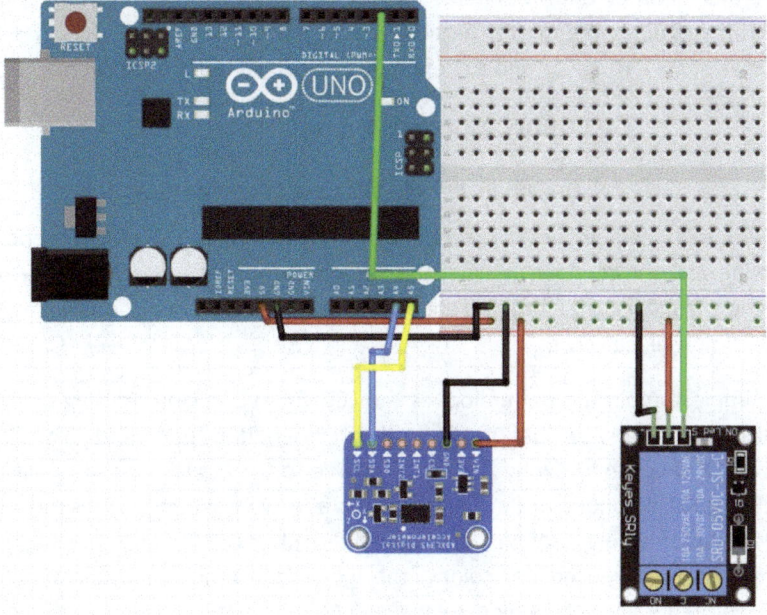

3.1.3 Digital electronics

In this chapter, we will deal with the basics of electronics, a main field of electrical engineering. In particular, we will take a closer look at digital electronics.

The basis of digital electronics is simple switching operations. The computer is one of the best examples of these switching operations and of digital electronics. The applications that a modern computer enables us to do are achieved thanks to switching operations performed by millions of transistors.

So, what is the basic principle behind a PC? Surely, you have heard this before. It is the so-called binary system, which is based on the two numbers "0" and "1". Communication in digital systems takes place with the help of these numbers, or with the help of various combinations of these two numbers.

Since the Arduino is basically nothing more than a very simple and stripped-down mini-PC, this principle is also applied here. The two binary numbers are mostly represented in today's electronic systems by the voltages 5V ("1" or HIGH value) and 0V ("0" or LOW value).

The restriction to only two numbers or voltage values seems to be very limiting, and it is very hard to imagine how a PC can achieve today's outstanding performance based on this system. However, this system and its simplicity makes sense. It simplifies the matter because it is extremely simple to recognize these two states, i.e., "0" or "1" and to definitely distinguish them from each other.

3.2 Important components in electronics

3.2.1 The diode and the light-emitting diode (LED)

A diode is a semiconductor component in electronics that has the property of allowing current to flow in only one direction (forward direction). The other direction is blocked for the current flow (reverse direction). You can imagine a diode simply like a valve.

The simplest application of a diode is the LED. The LED (light-emitting diode) is a semiconductor device that produces light when it is energized. The light is produced by current flowing from a DC source to the diode and through it. Since an LED is a semiconductor device, it also has a forward direction. This means that current can only flow through it in that direction. If an LED is connected incorrectly, no light will be produced. The color of the light and whether it is visible or not (e.g., infrared; generally determined by wavelength) is controlled by the doping and material used. Two major advantages of LEDs are: a) long life, b) low-power consumption. Compared to old-fashioned incandescent lamps, an LED can achieve a lifetime of several 10,000 hours and has many times better efficiency. Why is that? Conventional incandescent lamps produce an enormous amount of heat in addition to visible light, which means that the energy expended is converted not only into light, but primarily into heat. With LEDs, only a little heat is produced as a "waste or by-product" and almost all the energy can be used to produce the light. There are now different types of LEDs. The simplest design is shown in the following figure.

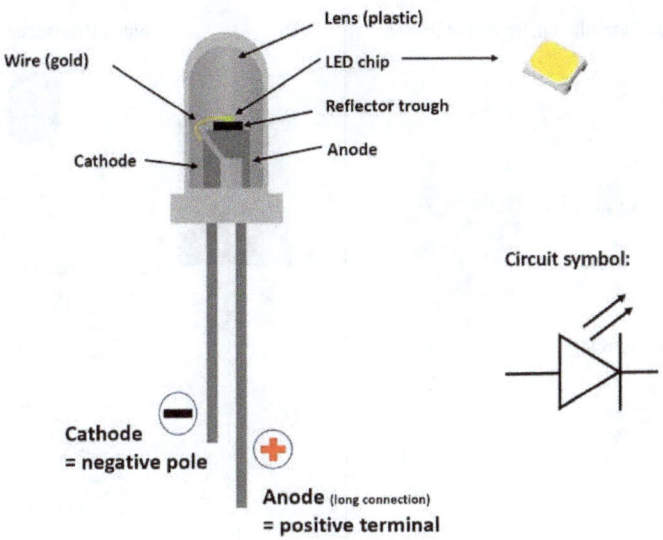

The heart and also the actual semiconductor element of the LED shown is the LED chip, which is placed on a reflector on the anode and emits the light. The circuit symbol of an LED consists of the diode circuit symbol with two additional slanted arrows, which are supposed to represent emitting light.

3.2.2 The transistor

A transistor is a simple three-terminal component that can best be thought of as a valve that controls the flow of water in a pump, for example. If we turn the control wheel of the valve in a certain direction, i.e., open, the water flow increases and if we turn it in the other direction, i.e., close, the flow decreases. The valve, in the case of the transistor, would be diodes and the water would be the current. Electronics in general, simplified, has a lot to do with switching elements and transistors also behave like switches. In addition to this switching capability, transistors also have the property of amplification, which would be equivalent to changing the valve ratio for the amount of water output. This amplification property is particularly important in the world of electronics. There are several types of transistors, one of the simplest being the bipolar transistor (**BJT**). Furthermore, there are for example the field effect transistor (**FET**) and the metal oxide semiconductor field effect transistor (**MOSFET**). All types of transistors have their special properties and are used in different applications.

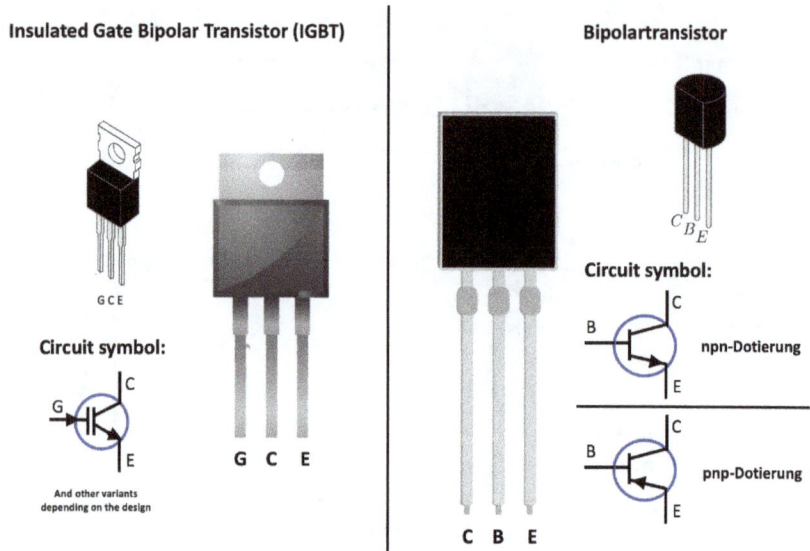

3.3.3 The capacitor

In simple terms, a capacitor consists of nothing more than two plates arranged parallel to each other and a dielectric between them. A dielectric is simply a weakly or non-conductive substance (solid, liquid, gas) with charge carriers that are <u>not</u> free to move. Capacitors are generally considered charge storage devices because, when an electrical potential is applied, they can store voltage (energy) in their plates.

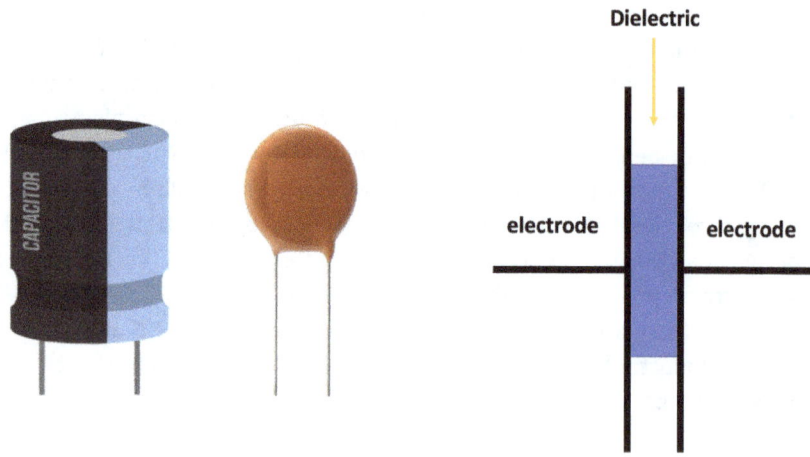

The ability of a capacitor to store charge depends on the area (A) of the plates, their spacing (d), and the dielectric (ε) between the plates. This ability to store charge is typically referred to as capacitance $\left(C = \frac{A}{\varepsilon d}\right)$, "ε" here is called the dielectric constant. As the charge in the plates increases, so does the voltage of the capacitor, and this continues until the capacitance is reached.

3.3.4 The resistance

Resistors are components that can be used primarily to - as the name suggests - apply resistance to something. In this case, the resistor acts against the current and can be used to limit the flow of current into a component that is connected to the resistor. Basically, every conductor (wire or similar) has a resistance that can be calculated depending on its length and cross-section. In our case, we use so-called sheet resistors (a design form of a resistor). Here, for example, there are the carbon film resistors and the metal or also metal oxide film resistors. With these types, the resistance value comes from a ceramic core with a layer of carbon or metal or metal-oxide. The resistance value can be measured either with the help of a multimeter or directly on the resistor by means of the colored rings. Each resistor has a color code consisting of 5 rings that reveal the resistance value. How to read this color coding must be explained in detail and would therefore go beyond the scope of this chapter. After all, we want to work with the Arduino as soon as possible. You can either look up this coding online, e.g., here: https://www.calculator.net/resistor-calculator or best, as said, use a multimeter to measure it.

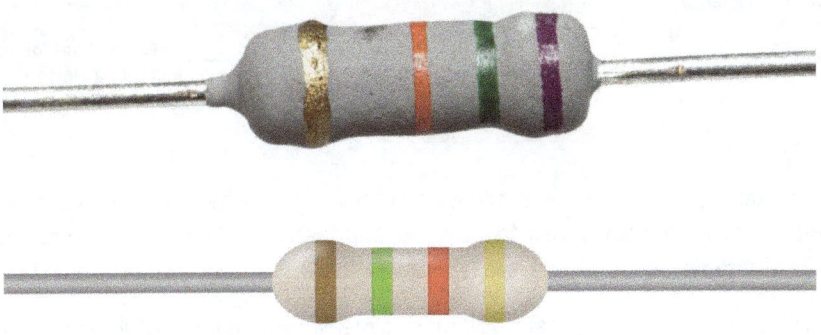

3.3 The multimeter: Current and voltage measurement in practice

In electrical engineering practice, multimeters are often used as measuring instruments. Multimeters with two terminals can measure voltage, current, resistance, capacitance, and inductance. But you can also measure the polarity of transistors and perform a continuity test with it. The continuity test tells us whether a circuit is shorted or not.

Multimeters can only measure one variable at a time (such as current or voltage). To measure multiple parameters, we would need to use several individual devices. The figure below shows a simple multimeter with the different measuring ranges. Depending on what you want to measure, you turn the dial to the appropriate range.

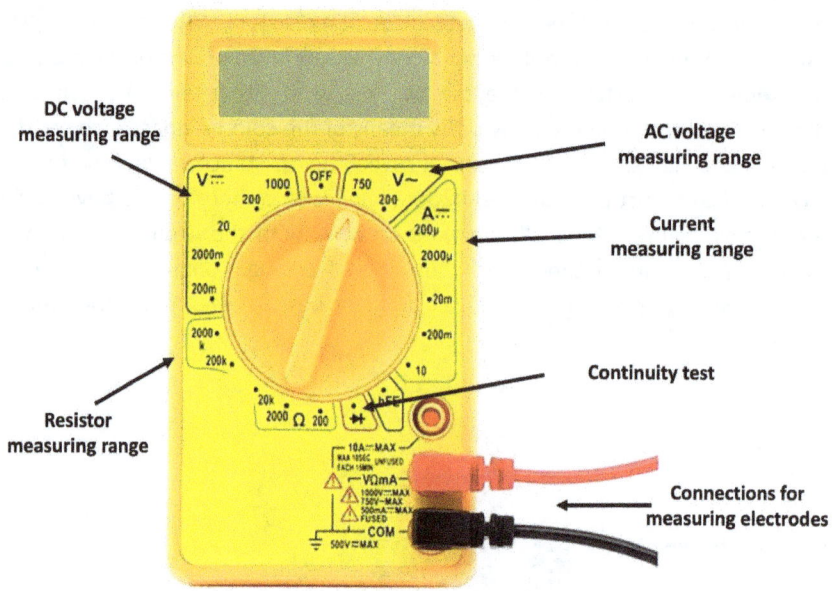

When taking a measurement, always start from the highest possible voltage or amperage, or the highest resistance value, and then turn the display setting down until a suitable value is displayed.

This means, for example, that if you make a measurement on a DC voltage source, and you suspect a value between 20 and 200 V, you first turn the setting range to 200 volts.

If you want to measure a voltage, you have to connect the measuring electrodes parallel to the voltage source or to the component you want to measure. As for a light bulb, for example, this would work like this:

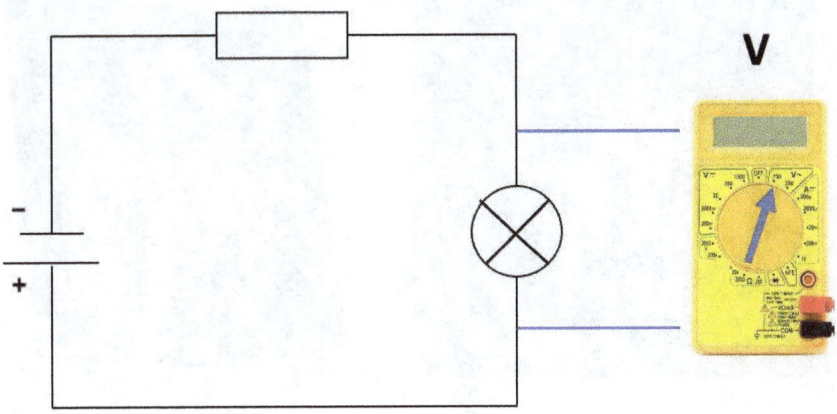

And if you want to measure the current of a consumer, you have to connect the measuring instrument (multimeter) in series to the consumer, i.e., disconnect the line. This would then work like this:

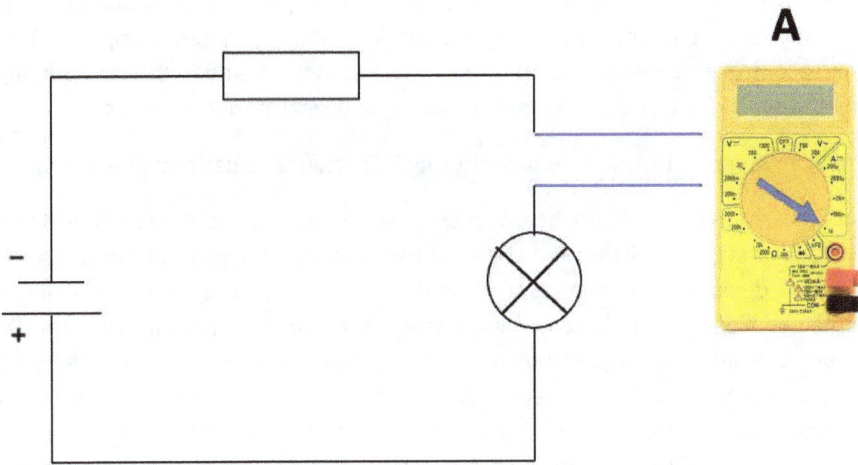

3.4 Embedded Systems

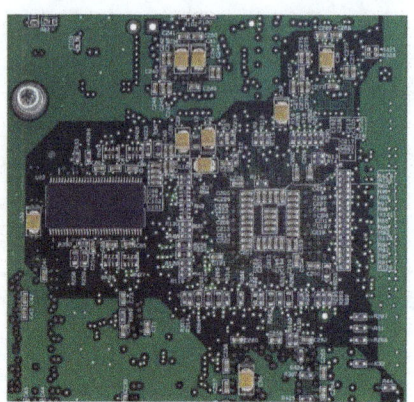 

Perhaps you have heard the term "embedded system" before. Perhaps you have also often wondered what it actually is and what it is used for.

In simple terms, an embedded system describes the presence of a type of computer in a technical system or on a circuit board (as in the case of the Arduino), that carries out signal transmission or data processing of input and output signals. This processing is done by a microcontroller, which is a tiny computer. This microcontroller is designed to perform certain functions and is basically nothing more than a tiny computer system consisting of a semiconductor chip.

You can program the microcontroller using PC software to perform operations.

As an example, in the figure below, you can see a system controlled by an Arduino UNO. This system switches the power of two devices (air conditioner and electric heater) depending on temperature and time. During off-peak hours (data is obtained from the utility company), it tries to match the electricity bill with the room temperature. A budget limit for the electricity price can be set with potentiometer RV1. It also attempts to turn the air conditioner on at night and off during working hours (6 days per week). In this circuit, the temperature is measured using LM35 temperature sensor and displayed on the LCD. In this complex system, the Arduino switches the air conditioner and heater by matching the temperature, time, and power bill (three feedbacks).

Arduino | Step by step

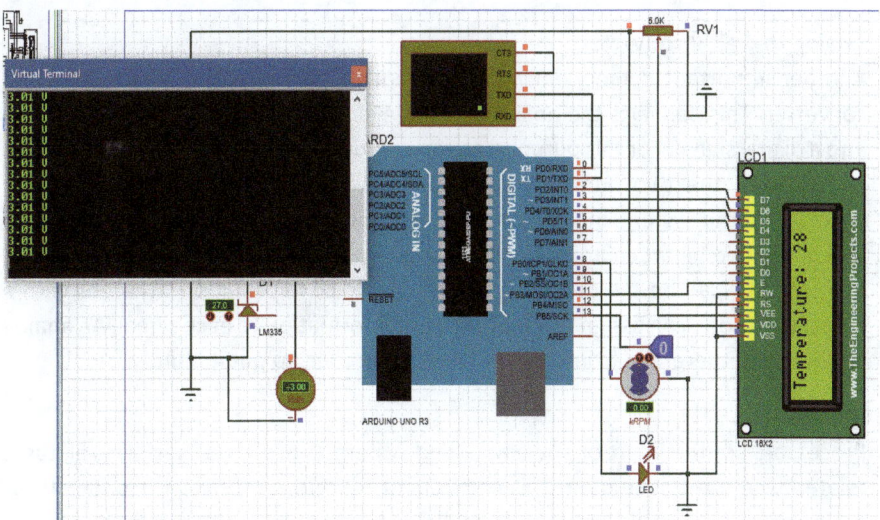

This example should only show what is possible with an Arduino, and that even complex systems are possible. Of course, we will start much simpler in one of the following chapters and learn everything step by step. So do not be afraid!

The system can be a stand-alone system, but it can also cooperate with other systems to perform a common task. In every embedded system there are circuits that perform the functions and send or receive instructions, i.e., transmit data in the form of voltage with the help of conductive elements.

In its simplest form, an embedded system consists of the following core components: Processor, sensor, actuator and an analog-to-digital converter as well as a digital-to-analog converter. We will look at these components in a little more detail below.

Sensor:
A sensor is a component that can convert physical changes in the real world into an electrical signal that can be used by a computer or electrical system to process data. Think of it like a human's sensory organs. With the help of eyes, ears and other sensory organs, our brain can interpret the outside world and thus create an image of it. And similarly, you can imagine it with computer systems. In this example, the sensors would stand for the sense organs and the microcontroller for the brain. The electrical signals coming from the sensors to the microcontroller allow the embedded system or the microcontroller to interpret what is happening in the outside world and then execute a response or program given by a programmer for a scenario by means of code.

Analog to digital converter:
Another important component of an embedded system is the analog-to-digital converter. This converts the analog signals (electrical pulses) sent by the sensors into a digital signal. For this purpose, as we already know by now, the binary system is used, i.e., the two numbers 1 and 0. These binary numbers represent the language of the system, in which a microcontroller can "understand" and "react". The difference between analog and digital signals is, among other things, that an analog signal can be the carrier of several pieces of information, whereas with a digital signal one can assign a unique piece of information to each signal. An analog signal would therefore "confuse" the microcontroller, to put it bluntly.

Processor:
Processors are the heart of any embedded system. A processor performs all tasks related to the received data. This component therefore receives the data, stores it, processes it and tells the system in what way it must react to this data.

Digital-to-analog converter:
A digital-to-analog converter is basically just the opposite of an analog-to-digital converter. It converts the digital signal sent by the microcontroller (which in turn is the response to the analog input signal converted to digital) back into an analog signal. So, why is the digital signal converted back to an analog signal? Simply because an analog signal can be understood by physical devices or actuators.

Actuator:
An actuator (e.g., an electric motor) converts the analog signal received from the microcontroller and the digital-to-analog converter into a physical action. There are mechanical, acoustic, chemical, thermal, and optical actuators that can perform physical actions in the real world according to their design. This is how embedded systems interact with the environment.

From this, there are a few steps we need to take when creating or designing an Arduino system.

1) First, we think about what **task** our embedded system, i.e., our Arduino, should perform (e.g., raise the blinds when the sun rises)
2) Then we think about which **sensors** we need for this (e.g., light sensor)
3) In addition, we need to think about the **program code** for this task (don't worry, we'll get to that)
4) And we still need an **actuator** to perform the task (e.g., a motor)
5) Of course, we also have to build the **board with** the **components** according to a previously considered circuit

Arduino | Step by step

4 The Arduino Board Hardware

In this chapter, let's take a look at the hardware, i.e., the board of the Arduino Uno. Each pin of an Arduino board is marked with a number or a label. The board works with 5V. In what follows, we will look a little more closely at the components of an Arduino, in this case the Arduino "Uno". Some of the most important components of the Arduino board are:

Digital pins can supply a 5V voltage or 0V. Similarly, these can also "detect" whether a voltage is present at a pin, and whether this is 5V or 0V. Logically, the latter is the case when there is no voltage. We can define in our program code whether a pin should be used as output or input. We will see how this works later.

Arduino | Step by step

An internal LED is connected to **pin 13 of** the Arduino. This LED can be useful in many situations. We will deal with it again later.

Another **LED** is connected to the **power pin to** indicate if the Arduino is receiving power.

The **AT mega microcontroller** controls the board, controls all input and output signals and serves as the digital control center of the Arduino. It is the processor of the board and later also contains the program code transferred by the user.

Five analog pins ("Analog in") are used to read analog voltage and convert it to digital voltage. This is done with the help of an analog-to-digital converter, which we have already learned about.

The two pins "GND" and "5V" are used to supply power to the circuits in a project. A **"3.3V"** power pin is also available. "GND" stands for "ground", which is the negative terminal of the board.

The board can be **powered** by USB cable or power plug. Arduino can work with voltages from 5 to 12V. Important: In no case should a higher voltage be applied!

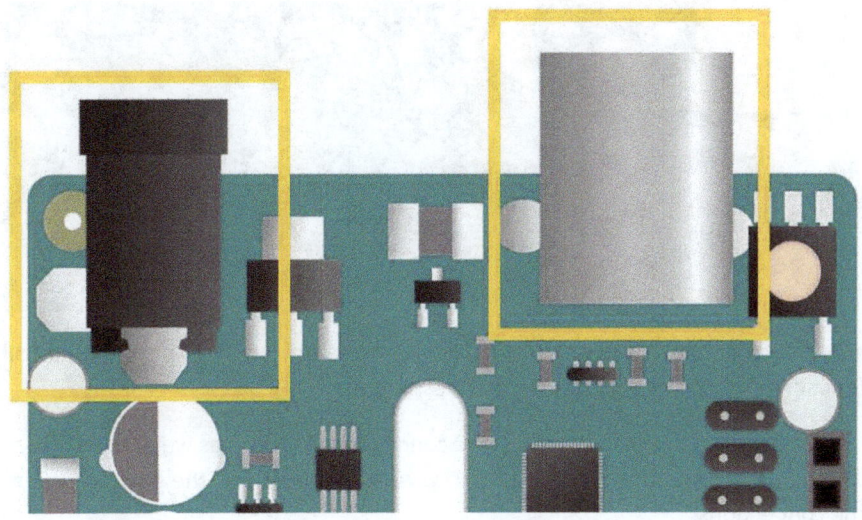

The two **pins** labeled **"TX" and "RX"** are connected to LEDs and indicate when communication is taking place, i.e., whether a signal is being processed or not, for example. This is especially essential for troubleshooting, which can be simplified considerably.

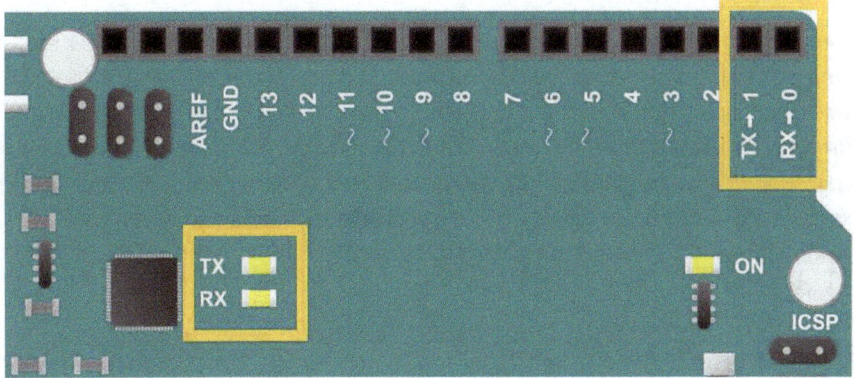

Finally, the Arduino can still communicate with the computer via a **USB port** (e.g., to transfer the program code to the processor).

You can reset the code at any time with a reset **button.** This button stops all functions that the board performs and restarts it.

We will neglect the other elements and connections for now, as we will come back to them later, or in the practical projects, we will understand the practical use of the connections.

In the next chapter, we turn away from the hardware and take a look at the Arduino software. Stay tuned, after the basics, exciting and great projects await us to implement.

Plug-in board / breadboard for expansion:

A breadboard is the best way to build a circuit as soon as it becomes a bit more complex or contains several parts. With a breadboard, there is an area for the power supply of the breadboard (+ and - imprint) and areas with letters and numbers. The pins that are in a row (letters: a-e and f-j; there is a non-conductive separation in between) are conductively connected to each other. This means, for example, h1 and i1 or h5 and i5 and j5 are conductively connected. The components and cables are plugged into the respective pins and thus connected to each other.

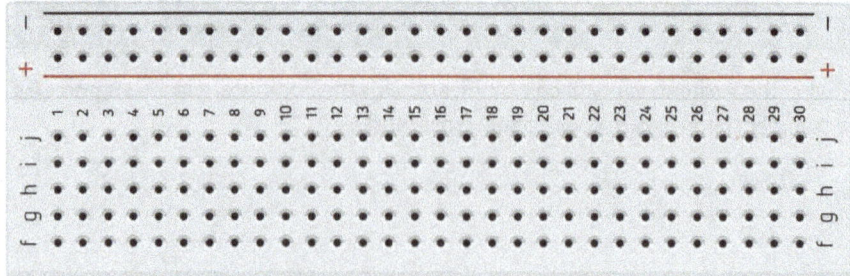

5 The Arduino Software (IDE)

The Arduino "Integrated Development Environment", commonly known as Arduino Software (IDE), consists of

- a text-based editor for writing lines of code,
- a news section,
- a toolbar,
- several menus.

The software can be connected to a microcontroller or the Arduino board to upload codes to run a program. Download the software here: https://www.arduino.cc/en/software. Below, we see a picture of the Arduino IDE environment:

```
void setup() {
  // put your setup code here, to run once:

}                                          ⟵ 1) setup-code

void loop() {
  // put your main code here, to run repeatedly:

}                                          ⟵ 2) main-code
```

In the following chapter, we will look at how to write a program code in this Arduino IDE, which you can later execute on the Arduino board. But first, let's take a look at an alternative method when programming an Arduino.

6 Writing a program code for the Arduino
6.1 Introduction to programming

Before we go into more detail about the Arduino IDE and learn how to create a program code in it, we will first get to know the method of block-based programming. This is simply an alternative to the (more complicated) text-based programming that follows later.

6.1.1 Block-based programming

Block-based programming is the simplest form of programming. This is mainly useful and great for people who have no experience with programming because you can achieve success quickly and easily. You can imagine it as if you put building blocks on top of each other in the software, each of which has a specific function. You just have to put these "building blocks" together in an orderly and meaningful way to get the finished code. For beginners, this type of programming is very useful to learn the basics of programming, methods and general operation.

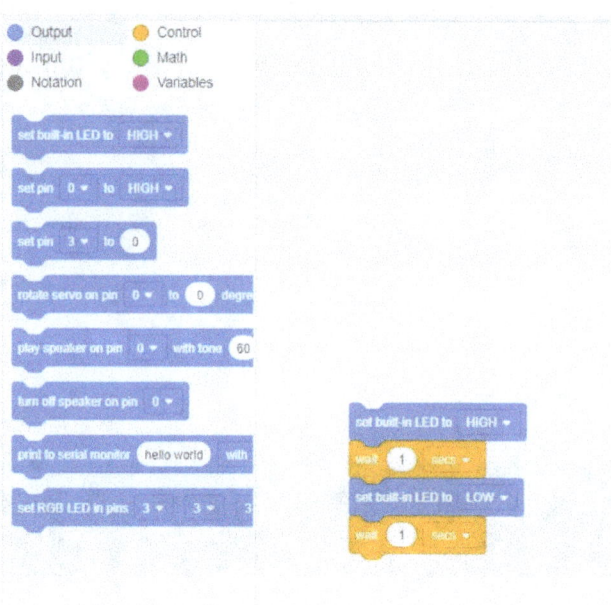

The best way to get started with block-based programming of an Arduino is to use Autodesk's TinkerCad software. TinkerCad is an online platform where, among other things, you can quickly and easily program an Arduino using the block-based programming we just presented. After creating an account at https://www.tinkercad.com/, you can get started. Through block-based programming, we mainly get the following three advantages:

1. as a beginner, we don't have to be afraid of small, but essential errors in the syntax (the program structure).

2. we can thus concentrate on the main task without worrying about the programming interface.

3. we can become familiar with the basic structure and flow of a text-based programming through the given blocks ("learning by doing").

Code blocks are divided into different categories. These categories are also color coded for better clarity. The following categories are available for selection:

"Output"

These blocks are used to instruct the actuators what to do (via the microcontroller). So, we control the output signals through this.

"Input"

With the help of these blocks, we bring the data from the sensors (i.e., the input signals) to the microcontrollers.

"Notation" (comments)

The blocks that can be found in this category do not directly affect the Arduino code but are used to indicate what the program code actually does. These blocks help the user to understand the program code.

"Control"

Control structures help to enable the microcontroller to make decisions based on the data it receives.

"Variables"

Variables are changing values that the program uses to execute mathematical functions or to store data.

When we use blocks from the different categories on TinkerCad, they align with each other like in a flowchart. But let's just take a look at a relatively simple example. For example, we would like to control an LED using block-based

Arduino | Step by step

programming and TinkerCad. In our example, we connect an LED to pin 2 (see picture) of the Arduino. Furthermore, we put a resistor between the negative pin of the LED and the negative terminal of the Arduino board ("GND") to control the amount of current flowing through it. This resistor will help us control the amount of current flowing through the LED and will prevent the LED from burning out.

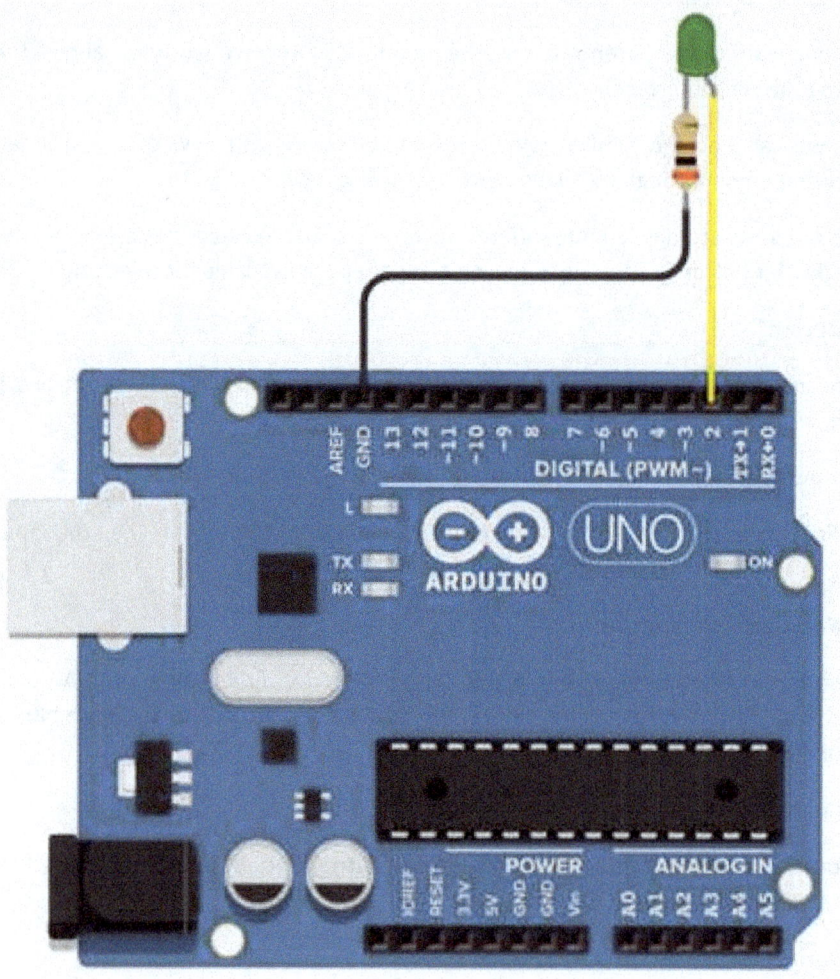

For example, if we now add the first block from the category "Output" in TinkerCad, as shown in the following figure, we can use it to turn on the LED.

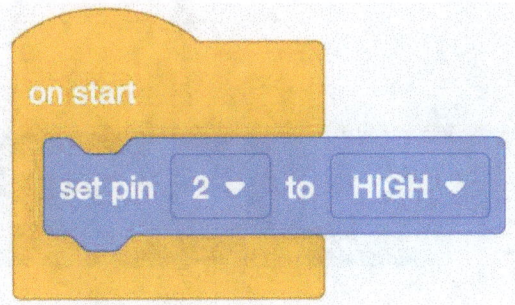

It also automatically implements the following circumstances in the program code (without us having to program them separately):

1. pin 2 is defined as output pin

2. pin 2 can no longer be used as input pin

3. in the code: "setup ()" the LED is connected to pin 2 of the Arduino

4. the actual switching on of the LED

Just play around with TinkerCad and the possibilities of block-based programming. That's the best way to learn. Meanwhile, we continue with the preparations for the - in this book mainly discussed - text-based method of programming.

6.1.2 Basics for using the Arduino IDE

In the Arduino software, you use the built-in text editor to write code for the Arduino. A code written with the Arduino software is called "Sketch". The editor includes the following functions, among others: Cut & Paste and Search & Replace. The message area provides the IDE's response when code is written. Such a response can also be an error message, for example. The console provides text-based output messages provided by the Arduino software (IDE) (e.g., general information, error messages). An Arduino code can then be saved with the file extension ".ino" when it is ready.

In the lower-right corner of the window, the configured Arduino board and serial number are displayed. The toolbar buttons give you options to review and upload programs, create "sketches", open and save, and open the serial monitor. Using the serial monitor, you can see what information the Arduino is sending to the PC (maps the communication between the Arduino and the PC).

In the next step, let's familiarize ourselves in detail with the elements of the program. For this purpose, let's take a look at the command bar with icons located in the upper part.

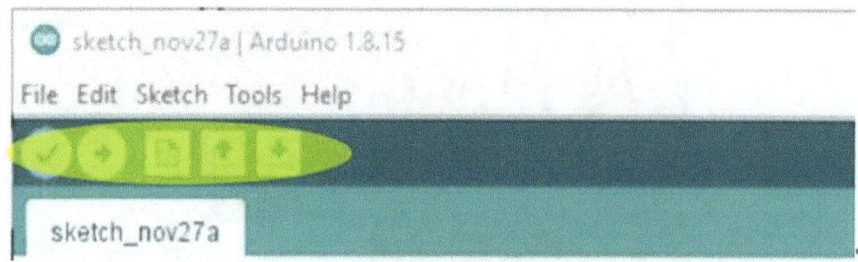

The small check mark is used to check the entered code for errors before compiling. Compiling means that you or the program translates the programming language into the machine language of the computer. Compiling then starts automatically with a click on the check mark.

With the arrow located to the right of the checkmark and pointing to the right, you can upload your code to the configured Arduino board (it must be connected via a USB port for this).

With the arrows pointing up and down, you can open or save a "Sketch". With the open arrow, you can also find example sketches.

The button next to it, which looks like a document, is used to create a new sketch.

And the small magnifying glass on the right side (picture above) of the program opens the serial monitor. This is used as already mentioned to monitor the communication between Arduino and PC / software or vice versa.

6.1.3 Libraries

Libraries represent an extension that allows us to give the Arduino additional functionality quickly and easily. It is basically nothing more than code that has already been written by eager members of the community. Especially for beginners and Arduino novices, this is a huge help in terms of time and effort. You can use a library by importing it. This is done in the Arduino Software IDE in the menu at the top under the item "Sketch". Here, you select "Include Library" and then select the library you want to use. You will get #include statements at the beginning of the code. Alternatively, you can just write them directly at the beginning of the code if you know the name of the library. The library manager is used to install new libraries in the "Sketch". To do this, open the program (IDE) and click on the "Sketch" menu and then on "Include Library" and then select "Manage Libraries".

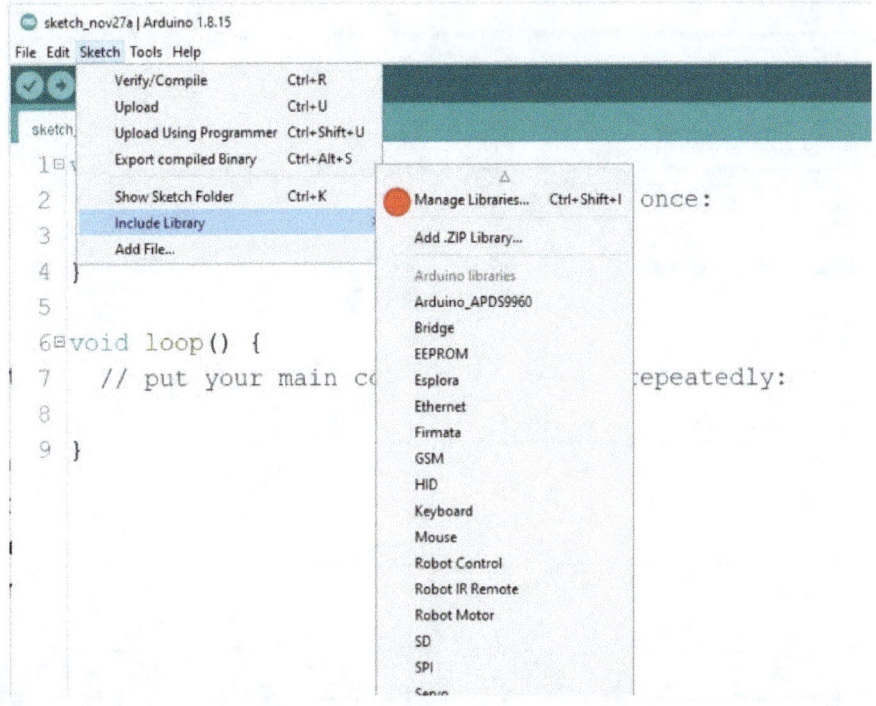

In the library manager, we then find a list of libraries that are already installed or ready for installation. Now we search for example once for the IMU library (still in the tab: "Manage Libraries"). To achieve this, we simply type in the abbreviation: IMU in the search field (position: top right). After that, we can select the version of the library. IMU is the abbreviation of "Inertial Measurement Unit" and is the name for a measurement unit / sensor network that is used to measure acceleration and rotation rates.

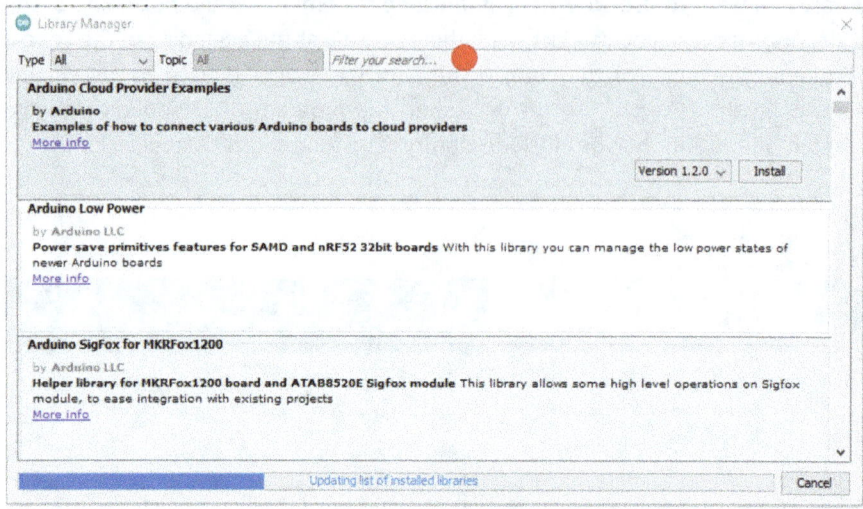

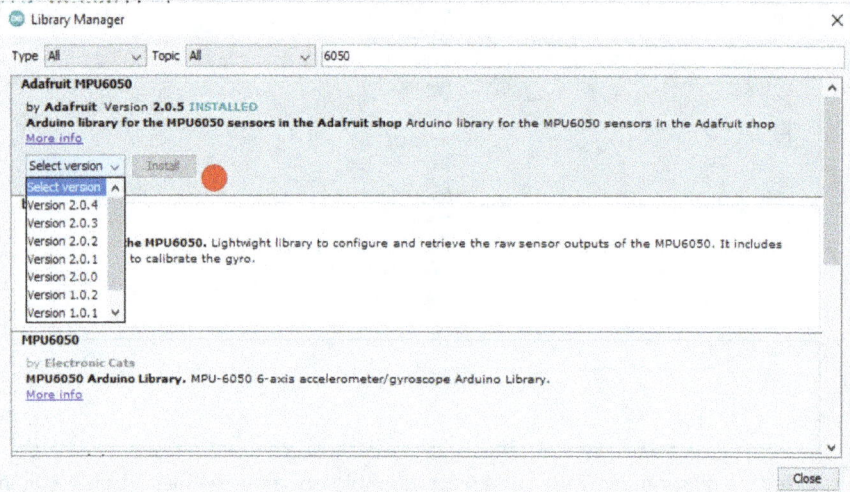

After we have selected the latest version, we can click on the "Install" button and then have to wait briefly for the new library to be installed. If we switch to the "Include Library" menu item again, we can check whether the library is now present, and the installation was therefore successful.

Manually import a library:

New or needed libraries can also be found online. These can be downloaded and installed as compressed "zip" files. Most libraries can be found on GitHub (github.com). GitHub is a management platform / community for software development. Once you have downloaded a library, you can load it into the program in the following way: In the Arduino IDE, go to "Sketch", then "Include Library", then select "Add .ZIP Library...".

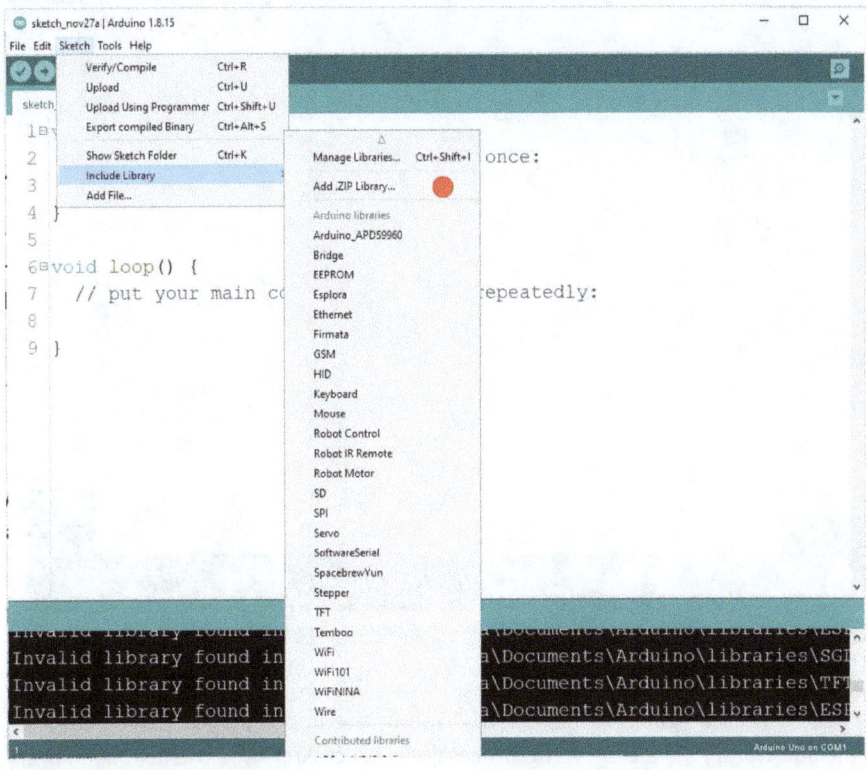

Then we are prompted to specify the location of the desired library. Navigate to the location of the downloaded file and select it.

I will tell you which libraries we need for our subsequent projects at the beginning of each project.

Arduino | Step by step

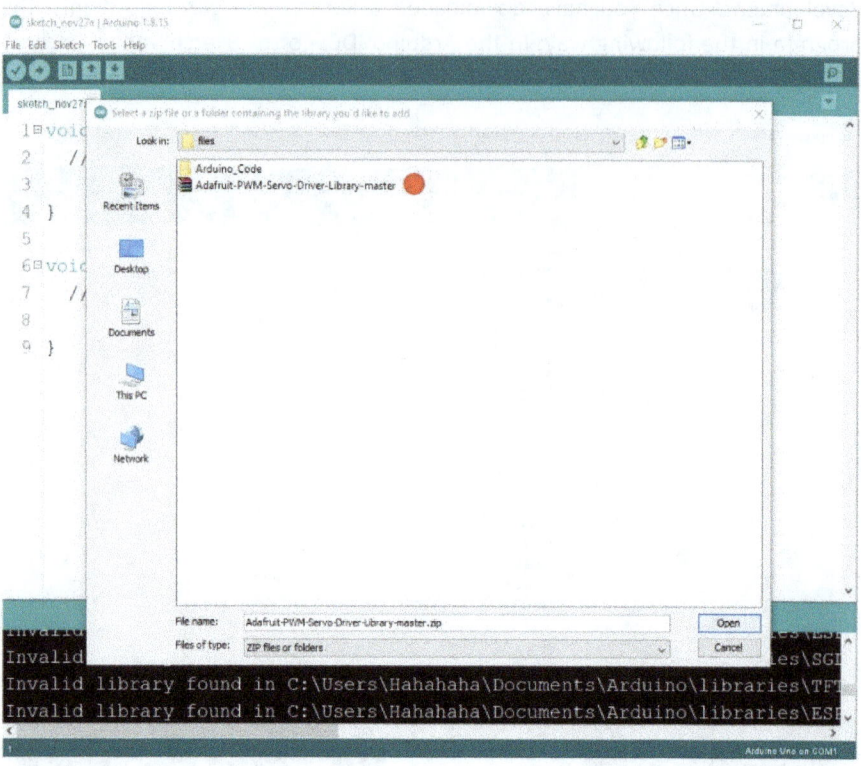

If we then click again on the "Sketch" tab in the upper menu bar and then on "Include Library", we can - if the process was successful - see the installed library in the lower area of the drop-down menu. Now the library is ready to be used.

I will tell you which libraries we need for our subsequent projects at the beginning of each project.

6.1.4 Serial monitor

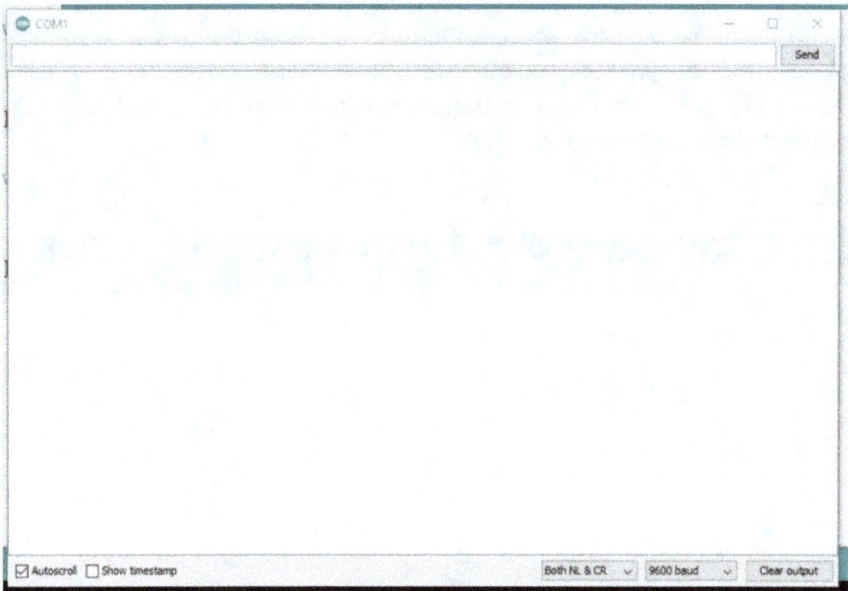

The serial monitor is used to display data sent from the Arduino to the computer. Here it is important to set the correct baud rate. Select the baud rate (bottom right) so that it matches the rate defined in the "Sketch" environment at "Serial.begin()". What this means exactly, you will understand better when we get to the practical projects. So, it's best to just read on first if you don't understand something right away. By the end of the book, it should be clear.

6.2 Basics for text-based Arduino programming

So that we do not have to work exclusively with block-based programming, we would like to get to know text-based Arduino programming in this chapter as well. This type of programming is a bit more difficult because we need to know the exact syntax and functions. Arduino code is written in the C++ programming language. Therefore, this chapter will give a basic overview of the structure of a text-based Arduino code, as well as introduce the most important functions, values, and structures. After working through this chapter, we also already get to the very practical programming and implementation of some great Arduino projects.

6.2.1 Structure of an Arduino code

We write the code in the Arduino IDE in the text-based editor in a so-called "Sketch". This contains the complete code, which is then transferred to the Arduino microcontroller with the Upload button (right arrow). Before that, you have to click the check mark to compile the code.

For any code written for an Arduino, there are two essential components:

The first of these is "setup ()". The code inside the following curly braces { } of this function is executed only once, and all relevant and essential information and structures for further code are listed here. For example, we tell the microcontroller here which pins are used as inputs and which pins are used as outputs.

The other essential component is "loop ()". The loop () function creates a loop. This means that the code inside the following curly brackets { } will be executed again and again, and not only once. Any task that the microcontroller is to perform is written into here. So, the basic code is written in here.

6.2.2 Syntax (program structure) for Arduino code

Let's first familiarize ourselves with the general program structure (syntax) when programming Arduino code. You can imagine the syntax, like the punctuation marks and paragraphs in a text. For example, after a sentence you make a period, but when programming an Arduino, you make a semicolon after each line of code. In addition, we must adhere to the following structure:

{ } Curly braces are used to start and stop a function. When the function is executed, the code inside these braces is executed.

; A semicolon tells the code that the current line of code is finished.

// Two slashes are used to write a comment to better understand as a human what the code is doing. All lines of code that start with these characters are ignored by the microcontroller.

/* A multiline comment can also be started with a slash followed by an asterisk. When you are done with this, you set the string in the opposite way, i.e., first an asterisk and then a slash ***/**. All lines of code between these characters are also ignored by the microcontroller.

With **#define** you can assign a name to a constant variable.

With **#include** you can include an external library into the code.

Practical tip: If a code should not work once, or the software gives an error when compiling, you should always check first whether you have used all syntax elements correctly. For example, check whether there is a ; after each line of code, or whether all comments start with **//**, or whether all the necessary brackets (open and closed) have been set.

6.2.3 Basic operators in Arduino programming

The following operators are used when programming in code to define logical commands when programming Arduino code:

== Two equal signs mean equality of two variables, e.g., x == y (x and y both equal).

!= An exclamation mark followed by an equal sign means unequal.

< means less than another variable (e.g., x < y; x is less than y).

> means greater than another variable.

<= less than or equal to another variable.

>= means greater than or equal to another variable.

% With a percent sign, you can get the remainder of a mathematical operation.

* An asterisk is used for multiplication.

+ is used for addition.

- is used for subtraction.

/ is used for the division.

= is used to assign a value to a variable.

&& operator for logical AND (value is true if both operators are true).

|| Operator for logical OR (value is true if one of the two operators is true).

* Value stored in the address used.

++ means add 1 to a variable.

-- means subtracts 1 from a variable.

+= Abbreviation for, e.g., x += y   → x = x + y.

-= Abbreviation for, e.g., x -= y   → x = x - y.

6.2.4 Basic data types for Arduino programming

array contains multiple values for one variable.

boolean stores the binary state of a variable (true or false).

byte stores a byte value.

char stores one character.

float stores a 4-byte value in decimal form.

double stores an 8-byte value; also in decimal form.

int stores a 4-byte number.

long stores an 8-byte number.

size_t stores the size of a variable in bytes.

string stores a text.

unsigned followed by e.g., **int** or **long** or other helps with negative numbers (an unsigned consideration is performed).

void is used for function declarations that do not return a value at the end of the function.

6.2.5 Constants and variables

When programming for an Arduino, data or values can be either a constant or a variable. Here is the difference:

Constants:

A constant is a fixed value, i.e., a data element to which a value has been permanently assigned.

The constant **HIGH** means that the microcontroller should apply 5V to a pin (additionally, which one has to be defined).

The constant **LOW**, on the other hand, means that Arduino should apply 0V to a pin (in addition, which one has to be defined).

The term **true** is used to define that a certain statement is true.

The term **false** is used to define that a particular statement is false.

INPUT defines that a pin (to be determined) is used for an input signal, i.e., that the microcontroller should read which voltage is present at this pin.

OUTPUT defines that a pin (to be determined) is used for an output signal, i.e., that the microcontroller should apply either 0V or 5V (HIGH or LOW) to this pin.

INPUT_PULLUP is used to connect an internal resistor to a pin.

LED_BUILTIN is used to control an LED connected to pin 13 of the Arduino.

int describes a fixed numeric value (integer).

Variables:

A variable is a data element in the Arduino program that associates a name or letter with an assigned value. Defining a variable is called declaring a variable in the programming language. You can perform all kinds of mathematical operations with a variable.

For example:

int x = 45;

Using this line of code, we declare an **integer variable that is** named **x** and has a value of **45**. When we have declared the variable in this way, we can use this variable in our program. Such a declaration must always take place first; otherwise the microcontroller will not know the variable.

Now we can calculate with this variable, for example and add the value 10 to this variable. This would then look like this:

x=x+10;

This line of code says that the new value for x is equal to the old value of x plus the constant 10.

We can also transfer the value from one variable to another. We do this by writing the new variable on the left side and the original variable on the right side.

For example, we want to store the value of x in a new variable, e.g., in the variable y. Then we have to write:

y=x;

Once a variable is declared, it is associated with the stored value throughout the program. If we try to assign this name to another data type, the IDE will give an error message.

Also, essential is the scope of the declared variable. This simply means that if we declare a variable at the beginning of the program, we can use it anywhere in the program. However, if we declare a variable only in a specific function, then the variable can be used only in that function.

In the following code, we can declare three variables for this purpose as an example and consider what the scope of these variables is.

```
int x=0;
void setup()
{}
void loop()
{
```

```
int y=10;
}
void new_function()
{
int z=15;
}
```

So now we have declared three variables named x, y and z. What about the scope?

x is a global variable and can be used in any of the functions (declared at the beginning of the program code), y was declared in the scope of **void loop ()**, so it can only be used in that scope, and z was declared in **void new_function ()**, so again it can only be used in that specific function. So, always pay attention to which variables you declare in which place.

6.2.5 Controlling the operation of the Arduino

To combine the individual variables, operators, and constants into a function or a working structure, we need expressions that create a control or command. The most important ones are as follows:

If is used for checking a condition and is used to perform an operation when that condition is met.

else is used for the action to be performed if the condition is not met.

else if is used when a second condition is to be checked if the first condition is not met.

break stops the code in a loop.

continue restarts the code in the loop.

while is used to create a small loop within a code. This is executed until a defined condition is met.

for is used to create a loop that is executed with a defined number of operations.

do while... is used to create a small loop that runs until a condition is met

goto makes the program continue in a certain line.

return will return a specific value at the end of the function.

6.2.6 Functions

Functions are basically nothing more than abbreviations for a code segment that you would actually have to write again and again for a certain action. Since some actions are needed frequently, it makes sense to bundle them in certain expressions - the functions. Functions are simply declared like variables.

In addition, functions bring some other advantages. Some advantages that functions offer, are:

- The code remains organized and structured.

- Debugging (which is troubleshooting if the code doesn't work) becomes straightforward.

- The code is efficient and clear.

- The code is straightforward to understand, even for new users.

As an example, below, we can create a function that adds two numbers:

```
int x=0;
int y=10;
int z=0;

void setup()
{}

void loop()
{
test_function();
}

void test_function()
{
z=x+y;
}
```

In this code, we have declared a function called **test_function**. At the beginning we used the word **void**, which means that the function does not return a value, but only performs the action, i.e., adds **x** and **y** and then stores them in **z**.

If we want to output the value for **z**, we need to construct the function as follows:

```
int x=0;
int y=10;
int z=0;

void setup()
{}

void loop()
{
int a=test_function();
}

int test_function()
{
z=x+y;
return z;
}
```

This function is of the integer data type. It adds **x** and **y**, then stores the result, i.e., the value, in **z** and then outputs the value of **z**, which is stored in an integer variable called **a**.

We have started the functions in both cases in the **void loop()** area. We did this because we want the function to be executed continuously, i.e., in a loop. If we wanted to execute the function only once at the beginning of the program, we would have put it in the **void setup()** scope.

Basic functions:

Some very basic and important, as well as already declared and thus ready to use functions utilized in Arduino programming are:

digitalRead () to read the digital input.

digitalWrite () to write to a digital output.

pinMode () to assign an order (make a pin connector of the board either an input or an output pin).

analogRead () to read the analog input.

analogWrite () to write to an analog output.

Some advanced functions

Stop every tone of a buzzer with **noTone()**.

To start a tone in a buzzer, we use **tone()**.

To read a pulse on a pin, we use **pulseIn()**.

pulseInLong() is used to read long pulses.

To shift a byte of data, we use **shiftIn()**.

To shift a byte of data, we use **shiftOut()**.

random() to find a random number within the bounds.

Time-related functions

To make the program wait for a certain time, we use **delay()**. The number we put in the brackets then describes the waiting time in milliseconds (0.001 s). E.g. : delay(1000) → code waits 1 second.

To make the program wait in microseconds (= 0.000001 second), we use **delayMicroseconds()**

To read the time that has passed since the program was started:

micros() (in microseconds) and **millis()** (in milliseconds)

Mathematical functions / operations

To get the absolute value of a number, we use **abs()**.

To specify constraints, we use **constrain()**.

To find the maximum of two numbers, we use **max()**.

To find the minimum of two numbers, we use **min()**.

To calculate the power of a number, we use **pow()**.

To find the square of a number, we use **sq()**.

To find the square root of a number, we use **sqrt()**.

Functions for working in bits and bytes

To calculate the value of a bit, we use **bit()**.

To set a specific bit to zero, we use **bitClear()**.

To read a single bit from a number, we use **bitRead()**.

To set a bit to 1, we use **bitSet()**.

To convert a number into bits, we use **bitWrite()**.

To get the leftmost byte of a number, we use **highByte()**.

To get the byte on the far right of a number, we use **lowByte()**.

Internal and external interruptions

To attach an external interrupt function to a pin, we use **attachInterrupt()**.

To remove an external interrupt function from a specific pin, we use **detachInterrupt()**.

To start the internal interrupts, we use **interrupts()**.

To stop the internal interrupts, we use **noInterrupts()**.

Functions used for conversion

byte() is used to convert a value to a byte.

char() is used to convert a value into a character variable.

float() is used to convert a value to a float variable.

int() is used to convert a value to an integer variable.

long() is used to convert a value to a long variable.

string() is used to convert a value into a string (text).

In the following, we will learn how to use some of these functions using sample projects. So don't worry if you don't know specifically how to use the functions, operators & conditions yet. Before we get to the projects, let's take a brief look at how to connect the board to the PC and load some program code onto the Arduino board.

6.3 Connect to the Arduino Board and Upload Code (Sketch)

To connect the Arduino Uno board with the PC or the Arduino IDE software, we first have to connect the Arduino board with a USB cable to the PC. Then we open the menu "Tools" in the Arduino IDE in the menu bar and select the correct board type - in our case the Arduino Uno - in the submenu "Board:" (see figure).

In the next step, we have to make sure that the correct USB port of the PC is assigned. We can also determine this under "Tools" in the "Port" submenu (see figure). This menu item can be found directly below "Board:". Here the port must be selected, which has "Arduino" or a similar designation, e.g., also "Genuino". The Arduino is connected to this port on your PC.

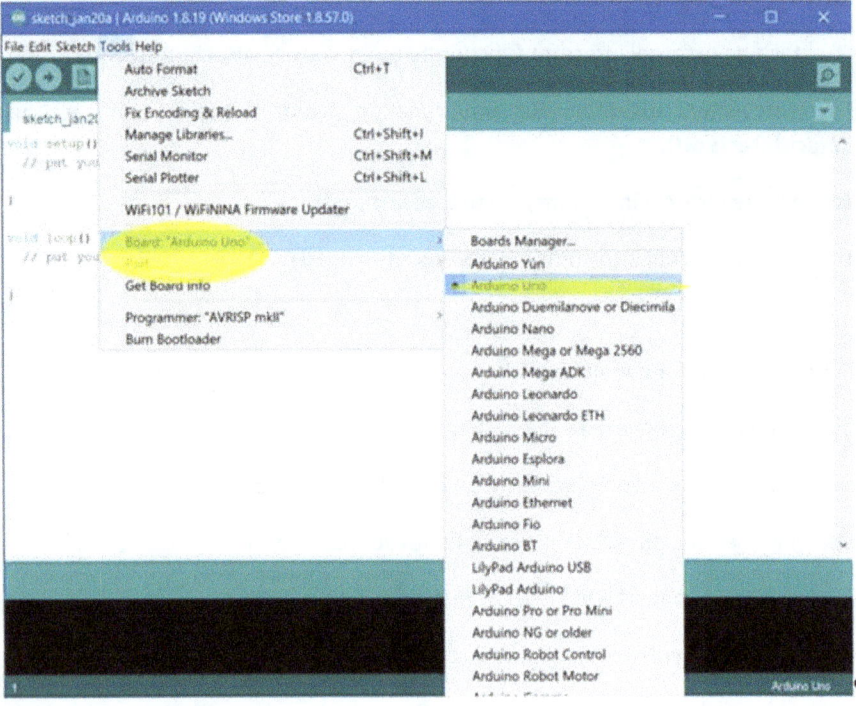

Now the Arduino board is properly connected to the PC or software, and we can start by writing the first program code, compiling it and loading it on the board. We do this as follows:

First, we write the program code or copy it into the text editor of the IDE (if you have a program code already complete, just delete the already existing syntax in the IDE). Then save the "sketch".

Then we press the small checkmark (menu bar above) to check and compile the code. If no errors were found, the message appears that the compilation was completed successfully. This may take some time depending on the code length.

Finally, we load the code onto the Arduino board by pressing the arrow pointing to the right (menu bar above). Then the Arduino will start executing the program code. Before that, you can open the serial monitor to monitor the communication between board and software.

7 Arduino DIY Projects
7.1 Project 1: A flashing LED and an SOS signal

In this project, we will control the state of a LED light. For this, we will use an Arduino Uno to switch the LED on and off with a delay of e.g., two (duration of the LED's glow) or one (switched off state) seconds.

Required components:

1x Arduino Uno Board

1x plug-in board (breadboard)

4x jumper wires

1x LED

1x 200 Ohm resistor

Wiring diagram:

We connect the LED to the Arduino using a resistor, the cables, and the breadboard as shown on the next page.

For this, we first supply the breadboard with power. We connect a red cable to the 5V pin of the Arduino board and the other end of the cable we plug into the breadboard as shown in the picture. We also connect a black cable to the GND pin of the Arduino board and the other end of the cable we plug into the breadboard as shown in the picture.

Then we place the LED (shorter leg of the LED to the resistor) and the resistor as shown and connect the LED with a black wire to the ground of the board. We need the resistor to limit the current. Here we use Ohm's law and the formula $R = U / I$. R stands for the resistance, U for the voltage and I for the current. Finally, we need a yellow cable (can also be another color), which goes from the LED to the board pin (2to be found at the digital pins above the Arduino logo).

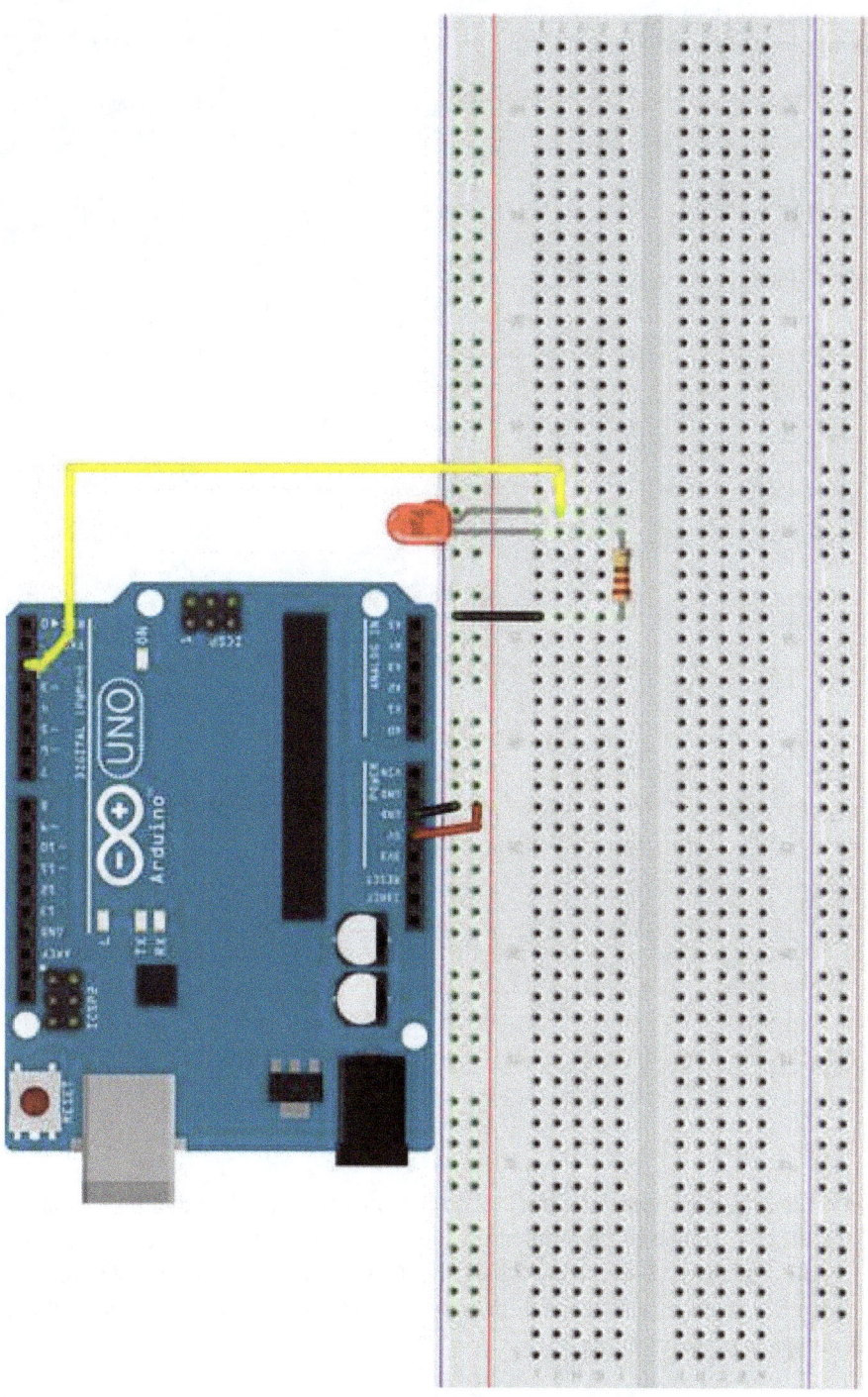

Program code for the Arduino IDE:

// We first declare the variable "led_pin" and assign it to the pin to which we have connected the LED (pin 2)

const int led_pin=2;

void setup() {

// Here we enter the setup code to be executed only once. We want the led_pin to be defined as an output pin, i.e., to receive an output signal so that the LED lights up, i.e. :

pinMode(led_pin,OUTPUT);
}

void loop() {

// Here we enter the main code to be executed repeatedly (in a loop)

digitalWrite(led_pin,HIGH); // first supply led_pin with 5V (switches LED on).
delay(2000); // then wait 2000 milliseconds (= 2 seconds) .
digitalWrite(led_pin,LOW); // then supply led_pin with 0V (turns LED off).
delay(1000); // then wait 1000 milliseconds (= 1 seconds)
}

// The loop then executes the program code repeatedly, i.e. the LED repeatedly switches off and on with the previously defined delay

// End of the program code. Tip: Check the correct number of { and }

Practice Task:

As an exercise, try switching the LED on and off at this point so that an SOS signal is sent! (SOS signal: three times short, three times long, three times short). Long signal = 2 seconds. Short signal = 1 second. Distance between short and long = 0.5 seconds. Distance of 5 seconds between several SOS signals *(You will find the solution on the next page).*

Solution:

We only have to change the part in **void loop()** in the above program code. For the SOS signal, this could look like this. A short signal is defined in the following, e.g., with 1 second, a long one with 2 seconds (LED on). Between the signals, there should be 0.5 seconds for disconnection (LED off).

```
void loop()
{

// The three short signals follow first:

digitalWrite(led_pin,HIGH); // first supply the led_pin with 5V (switches LED on)
delay(1000); // wait for short signal 1000 milliseconds (= seconds1)
digitalWrite(led_pin,LOW); // then supply led_pin with 0V (switches LED off)
delay(500); // wait for e.g. milliseconds500 to separate (= seconds0.5)

digitalWrite(led_pin,HIGH); // supply led_pin with 5V again (switches LED on)
delay(1000); // wait for short signal 1000 milliseconds (= 1 seconds)
digitalWrite(led_pin,LOW); // then supply led_pin with 0V (switches LED off)
delay(500); // wait for separation e.g. 500 milliseconds (= 0.5 seconds)

digitalWrite(led_pin,HIGH); // supply led_pin with 5V again (switches LED on)
delay(1000); // wait for short signal 1000 milliseconds (= 1 seconds)
digitalWrite(led_pin,LOW); // then supply led_pin with 0V (switches LED off)
delay(500); // wait for separation e.g. 500 milliseconds (= 0.5 seconds)

// then the three long signals follow:

digitalWrite(led_pin,HIGH); // first supply led_pin with 5V (switches LED on)
delay(2000); // wait for long signal 2000 milliseconds (= 2 seconds)
digitalWrite(led_pin,LOW); // then supply led_pin with 0V (switches LED off)
delay(500); // wait e.g. 500 milliseconds for separation (= 0.5 seconds)

digitalWrite(led_pin,HIGH); // first supply led_pin with 5V (switches LED on)
delay(2000); // wait for long signal 2000 milliseconds (= 2 seconds)
digitalWrite(led_pin,LOW); // then supply led_pin with 0V (switches LED off)
delay(500); // wait e.g. 500 milliseconds for separation (= 0.5 seconds)

digitalWrite(led_pin,HIGH); // first supply led_pin with 5V (switches LED on)
delay(2000); // wait for long signal 2000 milliseconds (= 2 seconds)
```

```
digitalWrite(led_pin,LOW); // then supply led_pin with 0V (switches LED off)
delay(500); // wait e.g. 500 milliseconds for separation (= 0.5 seconds)

// Finally, three short signals follow again:

digitalWrite(led_pin,HIGH); // first supply the led_pin with 5V (switches LED on)
delay(1000); // wait for short signal 1000 milliseconds (= 1 seconds)
digitalWrite(led_pin,LOW); // then supply the led_pin with 0V (switches LED off)
delay(500); // wait e.g. 500 milliseconds for separation (= 0.5 seconds)

digitalWrite(led_pin,HIGH); // supply led_pin with 5V again (switches LED on)
delay(1000); // wait for short signal 1000 milliseconds (= 1 seconds)
digitalWrite(led_pin,LOW); // then supply the led_pin with 0V (switches LED off)
delay(500); // wait for separation e.g. 500 milliseconds (= 0.5 seconds)

digitalWrite(led_pin,HIGH); // supply led_pin with 5V again (switches LED on)
delay(1000); // wait for short signal 1000 milliseconds (= 1 seconds)
digitalWrite(led_pin,LOW); // then supply the led_pin with 0V (switches LED off)

delay(5000); // to separate between several SOS signals e.g. wait 5000 milliseconds (= 5 seconds)

// the loop function then executes the SOS signal repeatedly and permanently

}
// End of the program code
```

Attention: When you try the code, don't forget the rest of the code structure as in the first project (identical). (Hint: Just replace the part inside "void loop ( )" from the first project with the code from here).

Excellent! We have successfully completed the first project! Let's move on to the second project. This one will be a bit more difficult.

7.2 Project 2: Temperature-based LED light

In this project, we will control the state of an RGB LED based on the temperature value detected by a temperature sensor. When the temperature is high, the light should turn red. On the other hand, when the temperature is low, the LED should glow purple / dark blue. When the temperature is optimal, the light should be green. In the areas in between, the LED color should change gradually. In a simulation (e.g., in Tinkercad) you can adjust the temperature by clicking on the sensor and moving the slider.

Required components:

1x Arduino Uno board

1x plug-in board

8x jumper wires

1x RGB LED

1x 200 Ohm resistor

1x TMP36 temperature sensor

Information on the LM35 temperature sensor:

The TMP36 provides an analog output voltage that is linearly proportional to the temperature in degrees Celsius. The measure range is from -40 °C to +125 °C. To convert the analog voltage into a temperature, a scaling factor of 10 mV/°C is required. The assignment of output voltage and temperature in Celsius can be read from the following linear function, e.g., 1V (signal of the sensor) = 50 degrees Celsius (see figure).

For the sensor to work, you have to connect its left pin to the positive pole and its right pin to the negative pole of a power source (2.7V - 5.5V). In this case, the flat side of the sensor head points to the front and the curved side to the back. The middle pin then provides the analog output voltage (which should be connected to an Arduino input pin).

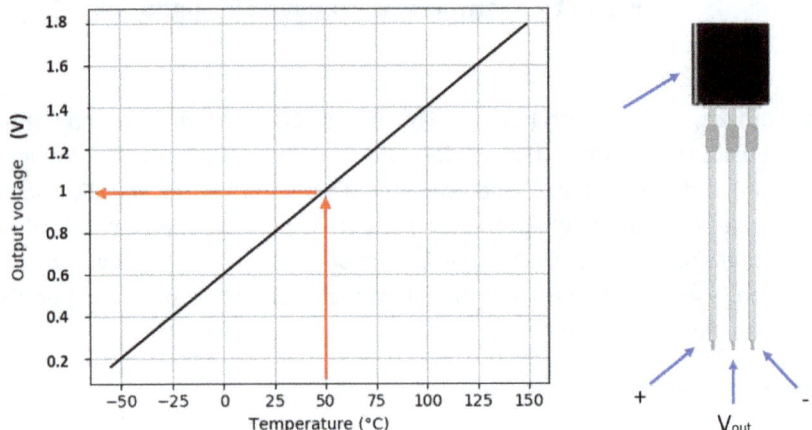

Information about the RGB LED:

An RGB LED can shine in three colors, namely red (r), green (g) and blue (b). The LED has two more connections than a normal LED and the color of the light depends on which connection is supplied with current. To control the LED, you need Pins 3, 6, 5, which are actually digital Pins that also allow pulse width modulation (PWM: pulsating voltage between 0V and 5V). You can recognize this by the small wave (also at Pins 9, 10, 11).

Wiring:

First we supply the breadboard with power again (see figure on the next page). For this, we connect a red cable to the 5V pin of the Arduino board and the other end of the cable we plug into the breadboard as shown on the picture on the next page. In addition, we connect a black cable to the GND pin of the Arduino board and the other end of the cable we also plug into the breadboard as shown in the picture on the next page.

Then we add the RGB LED and the temperature sensor (TMP). Make sure that each leg is correctly seated in the connectors. You need to be careful here not to break any of the legs when bending them into place. The two outer legs of the temperature sensor (TMP) are supplied with power by connecting a black and a red wire to the + and - pole of the breadboard. We connect the middle leg of the temperature sensor with a colored cable to the connector A0 of the Arduino.

We connect the legs of the RGB LED with colored cables to the digital pins 3, 5 and 6 of the Arduino board. Furthermore, we need a resistor to connect one leg of the RGB LED to the negative pole of the breadboard and thus also to the GND pole of the Arduino.

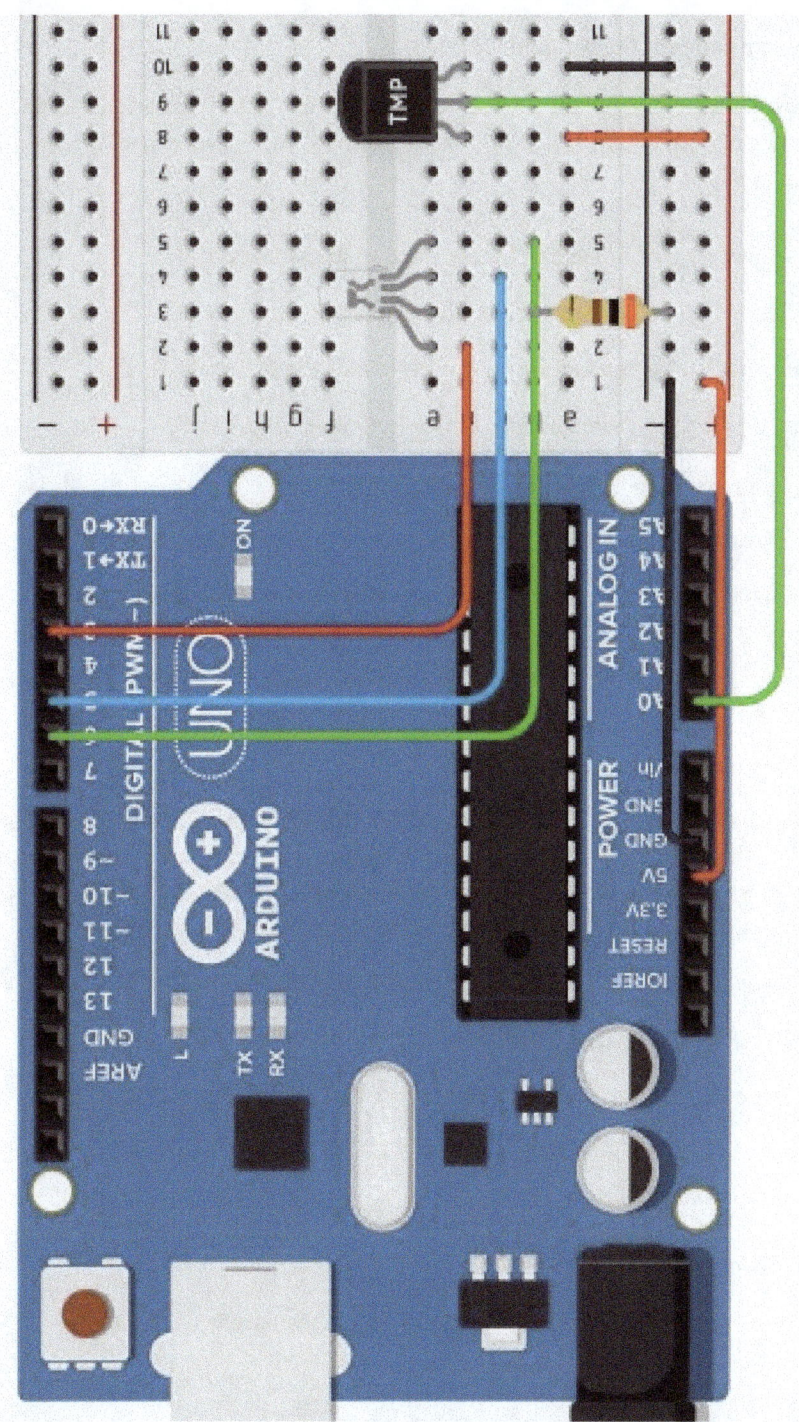

Program code for the Arduino IDE:

// We first declare our variables by assigning them to the respective connection pin. The abbreviations r, b, g, t stand for red, blue, green and temperature respectively.

const int r = 3; // Connection for red glow of the LED at pin 3

const int b = 5; // Connection for blue glow of LED at pin 5

const int g = 6; // Connection for green glow of the LED at pin 6

const int t = A0; // Temperature sensor is connected to pin A0

void setup()

// Here we enter the setup code to be executed only once. We want all pins connected to the LED to be defined as output pin and the pin connected to the temperature sensor to be defined as input pin.
{

  Serial.begin(9600);

// First we need the above command for the serial interface (data rate 9600 bit/s). This starts the communication between PC and Arduino board and the temperature is transmitted to the "serial monitor" in the IDE. Baud rate 9600. (Open serial monitor in the Arduino IDE to view measured values).

  pinMode(r, OUTPUT); // Definition of PIN r, i.e. pin 3 as output pin

  pinMode(b, OUTPUT); // Definition of the PIN b, i.e. the pin as 5output Pin

  pinMode(g, OUTPUT); // Definition of the PIN g, i.e. the pin as 6output Pin

  pinMode(t, INPUT); // Definition of PIN t, i.e. pin A0 as input pin

}

void loop()

// Here we enter the main code to be executed repeatedly (in a loop).
{

// First we define with the following code the function temp, which we will need in a moment. For this purpose the microcontroller is to read the value in PIN A0.

```
int sensorInput = analogRead(t);   // our TMP36 is an analog temperature sensor

double temp = (double)sensorInput / 1024; // determine percentage of the entered value

temp = temp * 5; // multiply by 5V to get the voltage

temp = temp - 0.5; // subtract offset (sensor has a 500 mV offset)

temp = temp * 100; // convert to degrees Celsius

Serial.println(temp); // shows us the temperature in the serial monitor (mV !)
```

// In the following we create "If" and several "Else if" conditions, which control the LED in several steps depending on the temperature of the sensor. Therefore we use "analogWrite" instead of "digitalwrite" (here only 0V or 5V would be possible; see first project). With the command "map" we define the temperature range or color range respectively.

```
int red_value  = map(temp, 20 , 125, 130, 255);

if(red_value<0){ red_value = 0;}

if(red_value>255){ red_value = 255;}

int green_value =  map(temp, 20, 30, 255, 100);


 if(temp > 20){

   green_value =  map(temp, 20, 40, 255, 100);}

 else{

   green_value =  map(temp, 10, 20, 80, 255);}


   if(green_value<0){ green_value = 0;}
```

```
if(green_value>255){ green_value = 255;}
int blue_value = map(temp, -40, 15, 255, 100);


 if(blue_value<0){ blue_value = 0;}
if(blue_value>255){ blue_value = 255;}


 analogWrite(r, red_value);
 analogWrite(b, blue_value);
 analogWrite(g, green_value);
 Serial.println(red_value);
}
```

// The loop then executes the program code repeatedly, i.e. the temperature is measured and interpreted continuously and the signal is passed to the LED.

// End of the program code. Tip: Check the correct number of { and }

7.3 Project 3: Light-dependent control of a motor (blind motor)

In this project, we want to control the blinds of a window with the help of a servo motor and an LDR sensor. This should happen depending on the amount of light coming in from outside. If we click on the photo sensor (LDR) in a simulation (e.g., in Tinkercad) and move the slider, the servo motor should move. In reality this movement should work depending on the incident light (e.g., sun).

Required components:

1x Arduino Uno

1x plug-in board

1x photoresistor (LDR sensor)

1x servo motor

9x jumper wires

1x Resistor (4.7 kOhm)

Things to know about photoresistance:

As the name suggests, a photoresistor can be thought of as a simple resistor that has the special feature of changing its resistance value depending on the amount of incident light. The less light falling on the sensor, the higher the resistance becomes. The more light falls on the sensor, the lower the resistance becomes (current can flow). The sensor is based on the photoelectric effect.

Wiring:

As usual, we first supply the breadboard with power. To do this, connect a black and a red cable from the GND pin and the 5V pin of the Arduino board to the positive and negative pole of the breadboard. By the way, it doesn't matter which GND pin of the board is used. Then we insert the photoresistor and the servo motor as shown in the picture. There is still one resistor missing and the rest of the wiring you can do as shown in the picture.

Arduino | Step by step

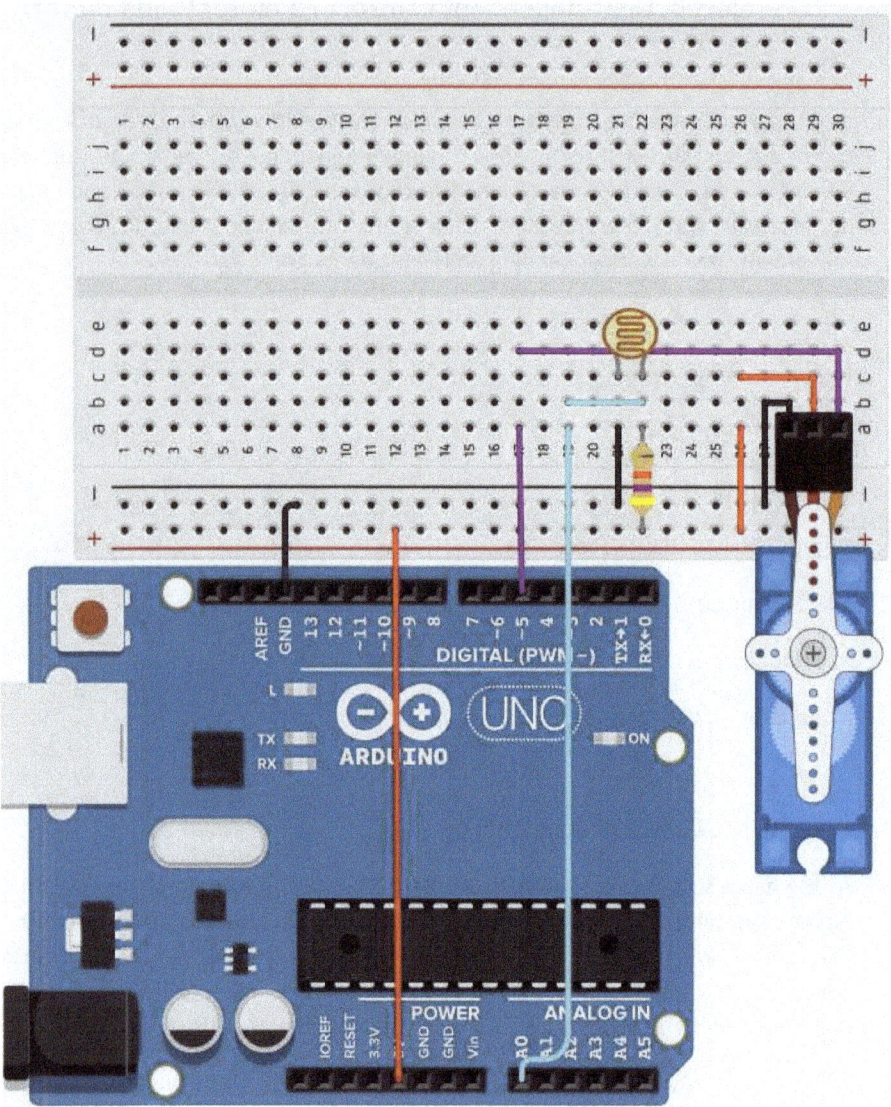

Program code for the Arduino IDE:

// First we include the required library for the servo motor in our program code. If the IDE gives an error message when compiling, you must first install this library via the library manager.

#include <Servo.h>

// Then we create a "Servo Object" so that we can control the servo motor.

Servo myservo;

// Then we declare the connection pins for servo motor and sensor (photoresistor)

const int servo_pin = 5;

const int sensor = A0;

// Then we have to declare variables for the position of the servo and for the properties of the photoresistor

int pos = 0; // Variable pos for saving the servo position

int light_pos =0; // variable to store the servo position at defined light

int max_light = 997; // This is the value we define as maximum light incidence

int intensity=0; // Light intensity at any position

void setup()

// Here we enter the setup code to be executed only once. Here we want to connect the servo with the "Servo object" and start the serial communication.

{

myservo.attach(servo_pin); // Connects the servo to pin with5 the servo object

  // Start serial communication. Set baud rate of the serial monitor:
  Serial.begin(9600);
  pinMode(sensor,INPUT);   // A0 is defined as input pin
}

void loop()
// Here we enter the main code to be executed repeatedly (in a loop).

{
  intensity = analogRead(sensor);   // reads the value of the LDR sensor (value between 0 and 1023)

  intensity = map(intensity, 0, 1023, 0, 180);   // scale value to use with servo (value between 0 and 180)

  myservo.write(intensity);   // sets the servo position according to the scaled value

  delay(100);
}

// End of the program code. Tip: Always pay attention to the ; at the end of the code

7.4 Project 4: Gas detection alarm

We will build a gas detector below that will sound an alarm if it detects a gas leak. The alarm will sound until the gas leak is stopped. In addition, LEDs are triggered depending on the amount of gas that the sensor detects. If there is a lot of gas leaking, all four LEDs should light up; if there is little gas, only one of the four LEDs should light up. This can also be checked in a simulation in Tinkercad.

Required components:

1x Arduino Uno

1x plug-in board

1x gas sensor

1x buzzer

14x Jumper wires

5x Resistor (1 kOhm) for LED and gas sensor

1x resistor (100 Ohm) for buzzer

4x LED

Wiring:

We connect all the components and the Arduino together on a breadboard as shown in the figure on the next page. Make sure that you use the correct pins. You can also arrange the components differently if you like. However, the circuit should remain comparable. But then you might have to change the variables or names in the following code.

Arduino | Step by step

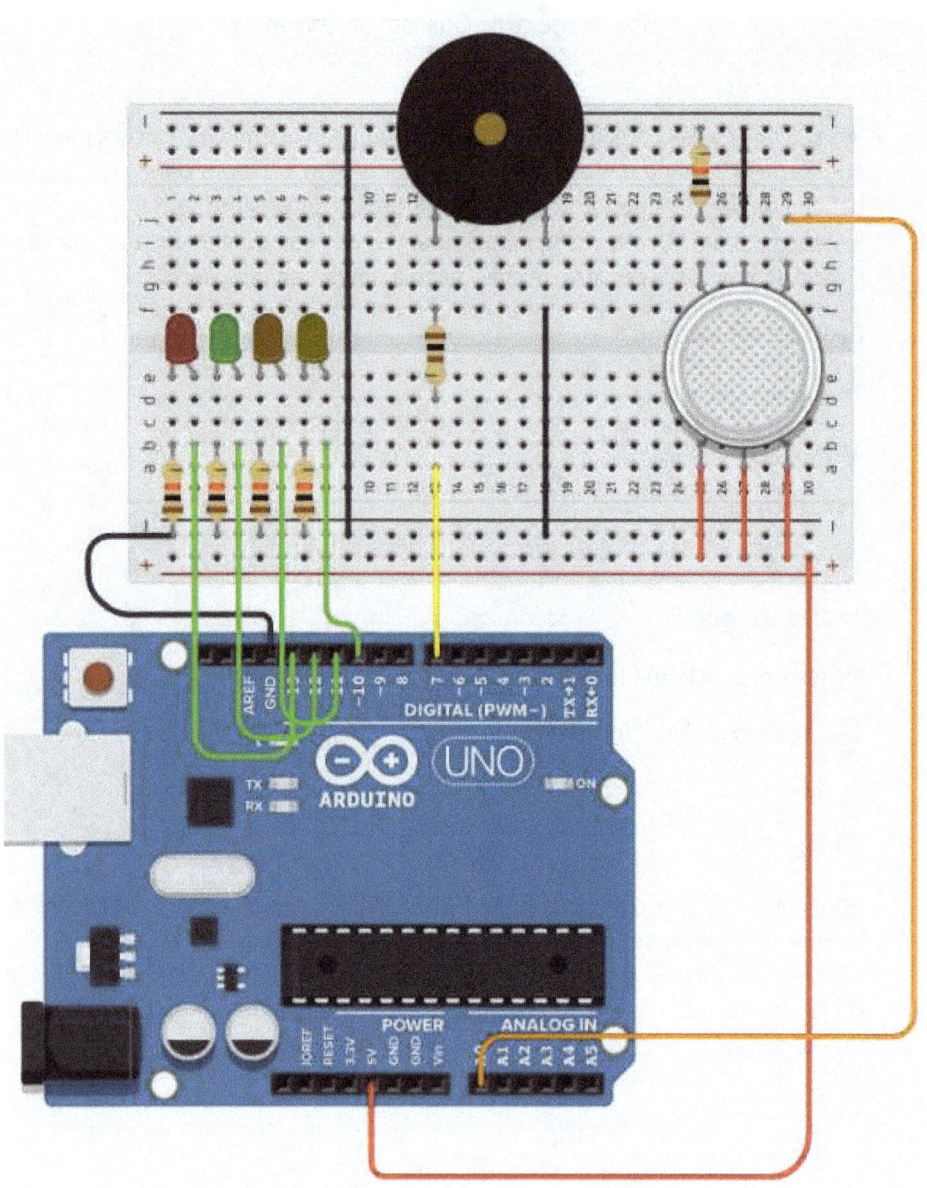

Program code for the Arduino IDE:

// We first declare our variable for the gas sensor by assigning the connector pin A0.

const int gas_s=A0;

void setup()

{

// Here we enter the setup code to be executed only once. We want PIN A0 which is connected to the gas sensor to be defined as input pin. Also we want all pins connected to the buzzer or LEDs to be defined as output pin.

   pinMode(gas_s, INPUT); // define gas sensor pin as input

   pinMode(7, OUTPUT); // define buzzer pin as output

   pinMode(13, OUTPUT); // define LED pin as output

   pinMode(12, OUTPUT); // define LED pin as output

   pinMode(11, OUTPUT); // define LED pin as output

   pinMode(10, OUTPUT); // define LED pin as output

// Then we start the communication with the serial interface (data rate 9600 bit/s) with the following code. This starts the communication between Arduino board and PC and the data is transferred to the "serial monitor" in the IDE.

   Serial.begin(9600);

}

void loop()

{

// Here we enter the main code to be executed repeatedly (in a loop).

// First we declare a variable that should read the sensor
        float gas_v=analogRead(gas_s); // float = floating point numbers

gas_v=(gas_v/373)*100; // Scaling the gas value

```
Serial.println(gas_v);
```

// In the following we create an "If" and several "Else if" conditions, which give us different control of the LEDs, and buzzer, depending on the gas value measured by the sensor. We use "digitalWrite" here.

```
if(gas_v>75) // If value is greater than 75, then...
{
// Buzzer einschalten
    digitalWrite(7,HIGH);
// switch on all LEDs
    digitalWrite(13,HIGH);
    digitalWrite(12,HIGH);
    digitalWrite(11,HIGH);
    digitalWrite(10,HIGH);
}
else if((gas_v>50) && (gas_v<=75)) // otherwise, if between 50 and 75, then...
{
// Buzzer einschalten
    digitalWrite(7,HIGH);
// switch on only 3 LEDs
    digitalWrite(13,LOW);
    digitalWrite(12,HIGH);
    digitalWrite(11,HIGH);
    digitalWrite(10,HIGH);
}
```

```
    else if((gas_v>25) && (gas_v<=50)) // otherwise, if between 25 and 50, then...
  {
// Buzzer einschalten
    digitalWrite(7,HIGH);
// switch on only 2 LEDs
    digitalWrite(13,LOW);
    digitalWrite(12,LOW);
    digitalWrite(11,HIGH);
    digitalWrite(10,HIGH);
  }
    else if((gas_v>0) && (gas_v<=25)) // otherwise, if between 0 and 25, then...
  {
// Buzzer ausschalten
    digitalWrite(7,LOW);
// switch on only 1 LED
    digitalWrite(13,LOW);
    digitalWrite(12,LOW);
    digitalWrite(11,LOW);
    digitalWrite(10,HIGH);
  }
}
// End of the program code
```

7.5 Project 5: Password protected mechanical system

In this project, we will create a system that is protected by a password. It will remain locked until the user enters the correct password. When the correct password is entered, the servo motor will move and open the system. We will assign different passwords to different users. Each user will have their own user ID and password. The system will only be unlocked if these two security features match and are correct. Moreover, after multiple incorrect entries, the red LED and buzzer shall be activated, and the entry shall be locked for 30 seconds. On the other hand, if the input is correct, the green LED should be activated. We can also replicate this project in Tinkercad (or in real). To unlock the system, we must first enter the user ID (e.g., #001, #002 or #003) and then confirm with the * key. It is best to also open the serial monitor to see if we have pressed the keys correctly. After that we can enter the respective password (user 1: 145278, user 2: 354691, user 3: 789541) and again confirm with the * key. Then the system should unlock after a short waiting time. Input thus e.g.: #001*, wait a short time, 145278*, system unlocks.

Required components:

1x Arduino Uno

1x plug-in board

1x keypad

1x LCD display

Various jumper wires

3x resistance (220 ohms each for LEDs and 250 ohms for display)

2x LED (green and red), 1x buzzer, 1x servo motor

Wiring:

We link all the components and the Arduino as shown in the figure on the next page. We don't use a breadboard this time because almost all the connections are between the Arduino and the individual components anyway. But you can also use a breadboard for this if you want to.

Arduino | Step by step

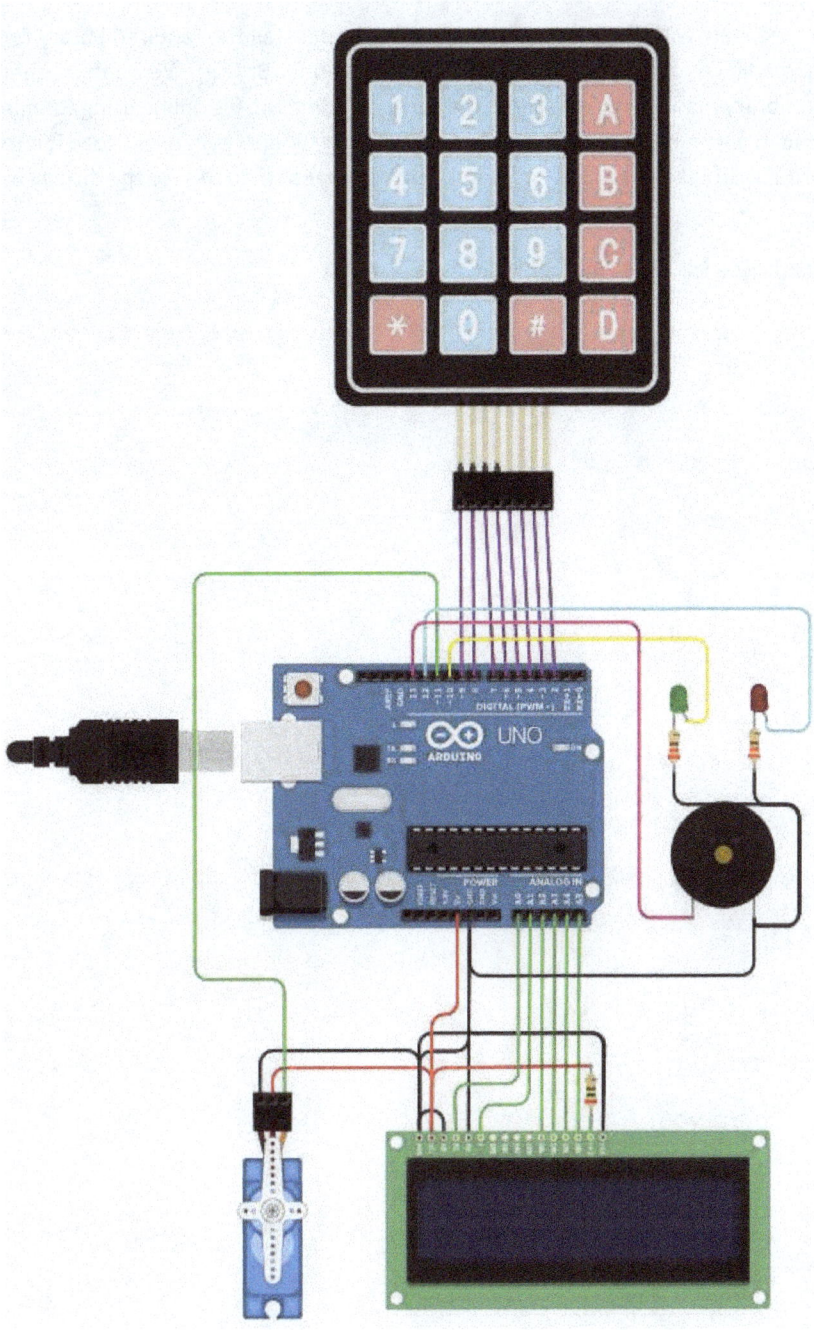

Before we start with the program code, we have to install the required library for the keypad ("Keypad.h" by Mark Stanley and Alexander Brevig). We do this either via the library manager or we search the corresponding zip. file online using Google and find it, e.g., here: https://playground.arduino.cc/Code/Keypad, and load it into the Arduino IDE (see previous chapters). You may need to do this for the display as well.

Program code for the Arduino IDE:

```
// First we include the necessary libraries for the servo motor, the keypad and the LCD display in our program code.

#include <LiquidCrystal.h>

#include <Keypad.h>

#include <Servo.h>


// Then we create a "Servo Object" so that we can control the servo motor.

Servo myservo;


// In the following we declare our variables for the green and red LED, as well as for the buzzer by assigning the connector pins 10, 12, 13.

const int gled=10;

const int rled=12;

const int buzzer=13;


// We then declare the number of rows and columns of our keyboard field (4 each)

const byte numRows = 4;

const byte numCols = 4;
```

// with "keymap" we define the keys on the keypad that can be pressed, according to the row and columns as they appear on the keyboard

```
char keymap[numRows][numCols] =
{
 {'1', '2', '3', 'A'},
 {'4', '5', '6', 'B'},
 {'7', '8', '9', 'C'},
 {'*', '0', '#', 'D'}
};
```

// Then we need code that maps the connectors of the keypad with the connectors on the Arduino

```
byte rowPins[numRows] = {9, 8, 7, 6}; // rows 0 to 3
byte colPins[numCols] = {5, 4, 3, 2}; // columns 0 to 3
```

// The following code initializes an instance of the "Keypad" class

```
Keypad myKeypad = Keypad(makeKeymap(keymap), rowPins, colPins, numRows, numCols);
```

// The following code creates a variable for the LCD display with the numbers of the pin interfaces assigned to the LCD display

```
LiquidCrystal lcd(A0, A1, A2, A3, A4, A5);
```

// In the following we assign the user IDs and passwords to:

```
String id1 = "#001";
String password1 = "145278";
String id2 = "#002";
String password2 = "354691";
```

```
String id3 = "#003";

String password3 = "789541";

// In addition, we declare the following variables:

int idcheck = 0;

int passcheck = 0;

int error = 0;

int pas = 0;

int ids = 0;

long prev = 0;

int idno = 0;

String input = "";

bool enter = false;

void setup()

// Here we enter the setup code to be executed only once.
{
  Serial.begin(9600); // start serial communication

  lcd.begin(16, 2); // Initialize LCD display: lcd.begin(columns, rows of display)

  delay(500); // wait 500 ms

  pinMode(gled,OUTPUT); // define green LED variable as output

  pinMode(rled,OUTPUT); // define red LED variable as output

  pinMode(buzzer,OUTPUT); // define buzzer as output

  lcd.setCursor(0, 0); // Set the position of the text on the display (column, row)

  lcd.println("ENTER THE ID"); // Display text : "Enter the ID" (display)

  lcd.setCursor(1 , 1); // Set the position of the text on the display (column, row)

  lcd.println("TO LOGIN"); // Display text : "To login" (display)
```

```
// Now we need code for the servo motor

myservo.attach(11); // The servo motor is connected to pin 11

  myservo.write(5); // The servo motor should move to position (5°)

  delay(1500); // wait 1500 ms

}

// End of the code for "void setup ()".

void loop()

// Here we enter the main code to be executed repeatedly (in a loop)
{

  // Below is the code that verifies the entered user ID:

get_char();

  if ( (idcheck == 0)&&( enter == true ))

  {

    enter = false;

    if ((input == id1) || (input == id2) || (input == id3))

    {

      if (input == id1)

      {

        idno = 1;

        Serial.print("Id no is ");

        Serial.println(idno);

      }

      else if (input == id2)

      {

        idno = 2;
```

```
      Serial.print("Id no is ");

      Serial.println(idno);

    }

    else if (input == id3)

    {

      idno = 3;

      Serial.print("Id no is ");

      Serial.println(idno);

    }

    idcheck = 1;

    error = 0;

    writingPassword();

    delay(2000);

  }

  else

  {

    wrongUser();

  }

  input = "";

}
```

// Below is the code that verifies the entered password and verifies and matches it with the previous entered ID:

```
if ( (idcheck == 1) && ( enter == true ))

{

  enter = false;
```

```
  if ((input == password1 && idno == 1) || (input == password2 && idno == 2)
|| (input == password3 && idno == 3))

    {

    Serial.println("succesfully login");

    passcheck = 1;

    error = 0;

    }
    else{

      wrongPassword();

    }

  input = "";

}
```

// **If the ID check was correct and the password check was also correct, the following should happen:**

```
 else if ( (idcheck == 1) && ( passcheck == 1 ))

 {

   successfulyLogin();

 }
```

// **If a wrong input is made several times, the input shall be locked for 30s and the red LED and the buzzer shall be activated. Below is the code for this:**

```
 if (error == 3)

 {

   wrong3times();

 }

}
```

```
void get_char(){

  char keypressed = myKeypad.getKey();

  if ( keypressed == '*' )

  { enter = true; Serial.println("enter"); return; }


  if (keypressed != NO_KEY)

  {

    input += char (keypressed);

    Serial.println(input);

  }
}
void wrongUser(){

  Serial.println("wrong user");

     idcheck = 0;

     error++;   // Increase the value of the "error" variable by 1
```

// The following message should appear on the display if the user ID is entered incorrectly:

```
     lcd.setCursor(0, 0); // Set position of the following text

     lcd.println("Wrong user"); // Display text

     lcd.setCursor(1 , 1); // Set position of the following text

     lcd.println("Try again"); // Display text

     delay(3000); // wait 3s
```

// allow input again:

```
        lcd.setCursor(0, 0); // Set position of the following text

        lcd.println("ENTER THE USER ID"); // Display text

        lcd.setCursor(1 , 1); // Set position of the following text

        lcd.println("TO OPEN LOCK"); // Display text

    }
void wrongPassword(){


  Serial.println("wrong password");

      idcheck = 0;

      error++; // Increase the value of the "error" variable by 1
// The following message should be shown on the display if the password was entered incorrectly:

      lcd.setCursor(0, 0); // Set position of the following text

      lcd.println("Wrong Password"); // Display text

      lcd.setCursor(1 , 1); // Set position of the following text

      lcd.println("Try again"); // Display text

      delay(3000); // wait 3s
// allow input again:

      lcd.setCursor(0, 0); // Set position of the following text

      lcd.println("ENTER USER ID"); // Display text

      lcd.setCursor(1 , 1); // Set position of the following text

      lcd.println("TO OPEN LOCK"); // Display text

    }
void writingPassword(){

  lcd.setCursor(0, 0); // Set position of the following text
```

```
    lcd.println("Correct Id"); // Display text

    lcd.setCursor(1 , 1); // Set position of the following text

    lcd.println("enter password"); // Display text

}
void successfulyLogin(){
```

// The following code will be executed if both the ID and password are entered correctly:

```
    digitalWrite(gled,HIGH); // green LED should light

    lcd.setCursor(0, 0); // Set position of the following text

    lcd.println("Correct Id and "); // Display text

    lcd.setCursor(1 , 1); // Set position of the following text

    lcd.println("password"); // Display text

    delay(3000); // wait 3s

    lcd.setCursor(0, 0); // Set position of the following text

    lcd.println("Door Unlocked"); // Display text

    lcd.setCursor(1 , 1); // Set position of the following text

    lcd.println("");

    myservo.write(180); // servo motor opens (180° movement)

   delay(1500); // wait 1.5s

}
```

// In the following we need the code for the case that a wrong input was made several times. Then, according to the project description, the input should be locked for 30s, the red LED should light up and a sound should be generated.

```
void wrong3times(){
```

```
lcd.setCursor(0, 0); // Set position of the following text
lcd.println("Wrong Password"); // Display text
lcd.setCursor(1 , 1); // Set position of the following text
lcd.println("wait 30 Seconds"); // Display text
error = 0;
idcheck = 0;
passcheck = 0;
digitalWrite(rled,HIGH); // red LED should light
digitalWrite(buzzer,HIGH); // generate sound

// wait 30s and display text
prev=millis();
while ((millis()-prev)<30000)
{
lcd.setCursor(1 , 1); // Set position of the following text
lcd.println("wait"); // Display text
lcd.println((millis()-prev)/1000); // Show remaining seconds
lcd.println(" Sec"); // Display text
}
digitalWrite(rled,LOW); // turn red LED off
digitalWrite(buzzer,LOW); // turn sound off
lcd.setCursor(0, 0); // Set position of the following text
lcd.println("ENTER THE CODE"); // Display text
lcd.setCursor(1 , 1); // Set position of the following text
```

lcd.println("TO OPEN"); // Display text

  }

// End of program code

7.6 Project 6: Remote unlocking mechanism

In this project, we will control a mechanism for opening and closing a gate with an IR remote control. To be able to open the gate, the code **16580863** should be entered on the remote control, for example. In addition, two RGB LEDs and a buzzer shall accompany the process. Furthermore, a temperature sensor should monitor the ambient temperature and issue an error message if the temperature is too high. Additionally, an LED shall be activated if the photosensor measures only little ambient light. Furthermore, we install a button / switch for manual operation.

Required components:

1x Arduino Uno

2x Plug-in board

1x IR remote control

1x IR receiver (sensor)

1x LCD display

Various jumper wires

6x Resistance

1x Potentiometer

1x buzzer

1x servo motor

2x RGB LED

1x DC motor

1x L293d motor driver

1x LDR sensor (photoresistor)

1x temperature sensor

1 x button / switch

Wiring:

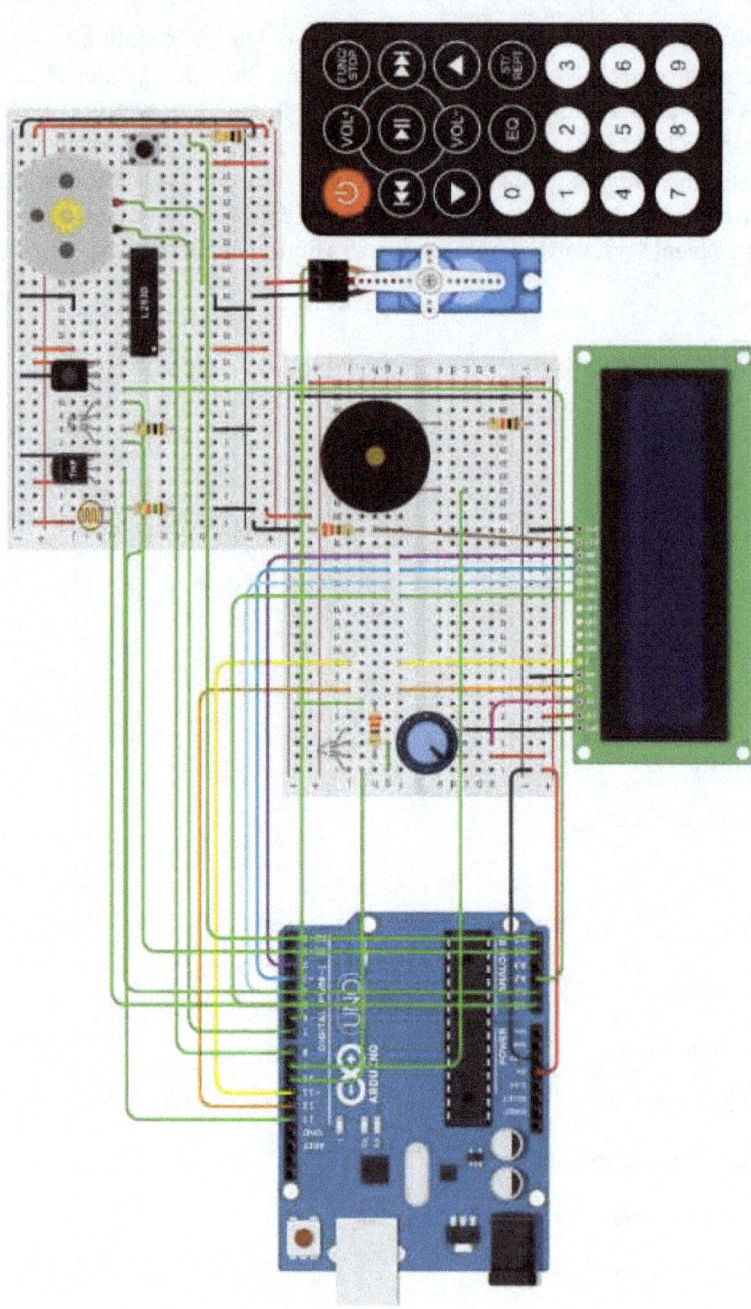

Before we start with the program code, we have to install the required library for the IR remote control ("IRremote.h" by Armin Joachimsmeyer). The best way to do this is to use the library manager, search for the library in the IDE and load it (see previous chapters). You may have to do this for the display as well.

Program code for the Arduino IDE:

// First we include the necessary libraries for the servo motor, the IR remote control and the LCD display in our program code.

#include <LiquidCrystal.h>

#include <IRremote.h>

#include <Servo.h>

// In the following we first declare the variable for the IR sensor (receiver) by assigning the connector pin A2.
int RECV_PIN = A2;

// We also need the following two expressions for the IR sensor (creates objects)

IRrecv irrecv(RECV_PIN);

decode_results results;

// Then we create again a "Servo object" so that we can also control the servo motor
Servo myservo;

int lightValue = 500;

// The following code creates a variable for the LCD display with the numbers of the pin interfaces assigned to the LCD display

LiquidCrystal lcd(12, 11, 5, 4, 3, 2);

// Below we define the function for opening:

void door_open()

{

  tone(8, 440, 100); *// tone at pin 8 with 220 Hz for 100 ms*

```
myservo.write(0); // The servo motor should move to position (0°)

delay(15); // wait 15ms

int temp=analogRead(A1); // declare variable "temp"; read temperature value

while(temp>250) // As long as the temperature is above 250 (25°, since 10mv = 1 °C), shall:

{

digitalWrite(13,HIGH); // Activate LED (RGB leg) at pin 13

digitalWrite(A4,LOW); // Disable LED (RGB leg) at pin A4

lcd.setCursor(0, 0); // Set position of the following text (LCD display)

lcd.print("Error temperature"); // Show text (LCD display)

lcd.setCursor(0, 1); // Set position of the following text (LCD display)

lcd.print("e above limit"); // Display text (LCD display)

temp=analogRead(A1); // read temperature value

}

digitalWrite(A4,HIGH); // Activate LED (RGB color) at pin A4

digitalWrite(13,LOW); // Deactivate LED (RGB color) at pin 13

digitalWrite(7,HIGH); // Activate pin 7 (motor)

digitalWrite(8,LOW); // Deactivate pin 8 (motor)

delay(3000); // wait 3000ms (=3s)

lcd.setCursor(0, 1); // Set position of the following text (LCD display)

lcd.print("Closing Door"); // Display text (LCD display)

digitalWrite(7,LOW); // Deactivate pin 7 (motor)

digitalWrite(8,HIGH); //Enable pin 8 (motor)

delay(3000); // wait 3000ms (=3s)

digitalWrite(7,LOW); // Deactivate pin 7 (motor)
```

```
  digitalWrite(8,LOW); //Enable pin 8 (motor)

  delay(100); // wait 100ms

  myservo.write(90); // The servo motor should move into position (90°)

  delay(15); // wait 15ms

}
```

// In the following we define the function for the LED when there is too little light:

```
void out_door_light()

{

  int light=analogRead(A0); // Declare variable "light"; read light value

  if (light>lightValue) // If light quantity is500 over, then:

  {

    digitalWrite(10,LOW); // Deactivate pin 10 (second LED)

    lcd.setCursor(0, 0); // Set position of the following text (LCD display)

  lcd.print("Light=OFF"); // Display text (LCD display)

  }

  if (light<lightValue) // If light quantity is500 under then:

  {

    digitalWrite(10,HIGH); // Activate pin 10 (second LED)

    lcd.setCursor(0, 0); // Set position of the following text (LCD display)

  lcd.print("Light=ON "); // Display text (LCD display)

  }

}


void setup()
```

```
// Here we enter the setup code to be executed only once.
{
// Initialize LCD display: lcd.begin(columns, rows of the display)
  lcd.begin(16, 2);
// Start the IR sensor with the following code
  irrecv.enableIRIn();
// The servo motor is connected6 to pin
  myservo.attach(6);
// The servo motor should move in position 90° degrees
  myservo.write(90);
  delay(15); // wait 15 ms
  digitalWrite(A4,HIGH); // Activate LED (RGB leg) at pin A4
  digitalWrite(13,LOW); // Deactivate LED (RGB leg) at pin 13
}
void loop()
// Here we enter the main code to be executed repeatedly (in a loop)
{
  out_door_light();
  if (digitalRead(A5)==HIGH) // If terminal A5 gets current (5V, because HIGH), then:
  {
    lcd.setCursor(0, 1); // Set position of the following text (LCD display)
      lcd.print("opening Door"); // Show text (LCD display)
      door_open(); // Execute the open function
    lcd.setCursor(0, 1); // Set position of the following text
      lcd.print("Door Open"); // Show text (LCD display)
```

```
    }
else{
    lcd.setCursor(0, 1); // Set position of the following text
    lcd.print("closing Door"); // Display text
    myservo.write(0); // Run function for closing
    lcd.setCursor(0, 1); // Set position of the following text
    lcd.print("Door closed"); // Display text
    }
```

// The following code says the following: If a signal was given by the IR remote control or received via the IR sensor, and if the signal corresponds to the following code "16580863":

```
   if (irrecv.decode(& results)) // Was an IR signal received?
   {
    if (results.value == 16580863) // does signal match code?
    {
     lcd.setCursor(0, 1); // Set position of the following text (LCD display)
      lcd.print("opening Door"); // Show text (LCD display)
      door_open(); // Execute the open function
     }
     irrecv.resume(); // Receive next value
      lcd.setCursor(0, 1); // Set position of the following text (LCD display)
      lcd.print("Door Open"); // Show text (LCD display)
    }
    delay(100); // wait 100ms
   }
// Ende des Programmcodes
```

Closing words

Excellent!

You've done it, you've worked through the beginner course.

Congratulations!

In this book, I have tried to bring you the basic knowledge for the use of an Arduino simply explained closer. I hope that I have succeeded to some extent and that this book has brought you a well understandable and practical introduction to the world of the mini-PC, and you now understand why the Arduino is such a great system and what you can do with it!

The aim of this book was to give you an understanding of how electrical engineering and programming accompanies us in everyday life, and the basic principles involved. It should be a book that creates an understanding of the theoretical background knowledge and practical application.

With this basic course, you should now know everything you need to know to use an Arduino as a beginner! Of course, it makes sense not to stop at this point and rather look into an advanced book to learn even more about creating systems using an Arduino.

Together, we have accomplished quite a bit in this course! Be justifiably proud of yourself if you made it to the end!

If you liked this book, I would be pleased if you leave me a rating and short feedback, as well as recommend the book! Thank you very much.

One final tip:

If you ever get stuck, take a look at the following site, where you will find many and great learning materials about the Arduino:

https://www.arduino.cc/reference/en/

If you are also interested in other books of mine on similar topics, be sure to take another look at the next pages.

Thank you very much!

Arduino | Step by step

Books on topics you might also like

All books are available online on the usual sales platforms. It's best to just search for the title, or feel free to visit my author page. Some of the books may not be published yet and will be released or found soon. Take a look at the books of your choice and your copy as e-book or paperback!

3D Printing:

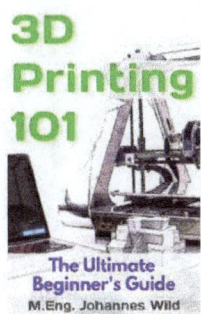

CAD, FEM, CAM (3D Object Creation, Design, Simulation):

Electrical Engineering:

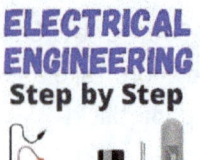
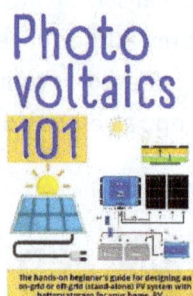
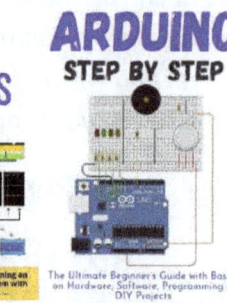

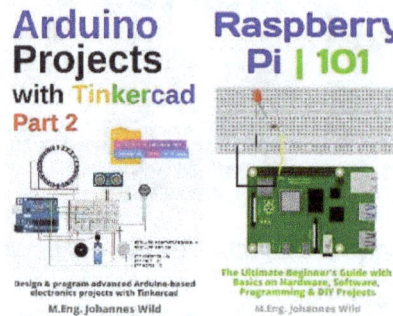

Programming and other Software:

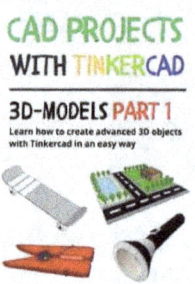

There are also identical video courses for some of these books:

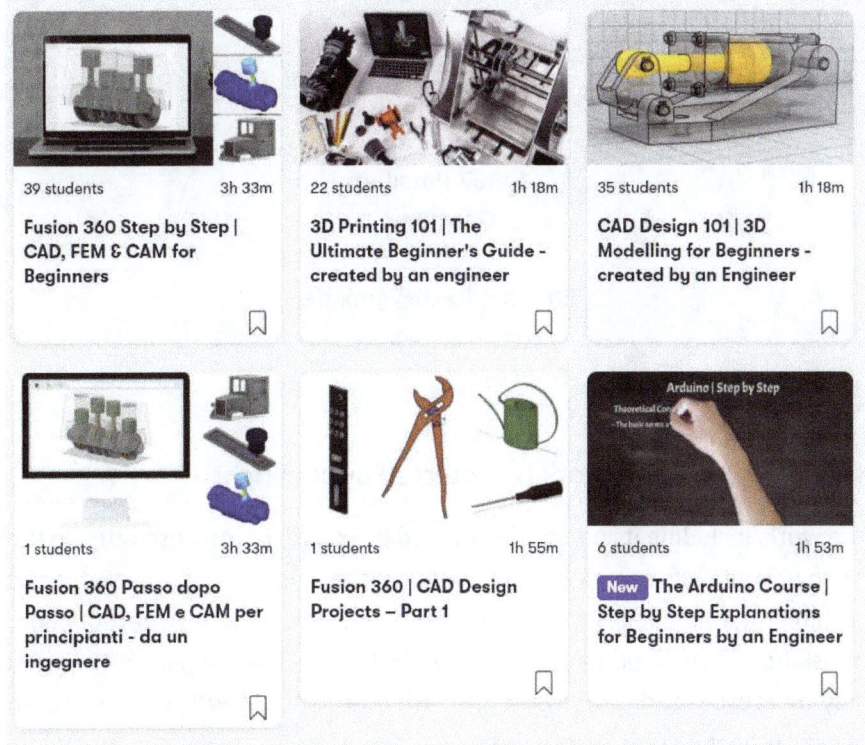

They are hosted on the learning website: skillshare.com

Be sure to use my following friends & family referral link to get a month of membership for free!

(I will get a little bonus if you choose to stay, so we will be both happy. Thanks in advance!)

https://www.skillshare.com/r/profile/Johannes-Wild/854541251

It is best to copy the link in your browser to access the free month!

Sign up today and deepen your knowledge!

Imprint of the author / publisher

© 2023

Johannes Wild
c/o RA Matutis
Berliner Straße 57
14467 Potsdam
Germany

Email: 3dtech@gmx.de

This work is protected by copyright

The work, including its parts, is protected by copyright. Any use outside the narrow limits of copyright law without the consent of the author is prohibited. This applies in particular to electronic or other reproduction, translation, distribution and making publicly available. No part of the work may be reproduced, processed or distributed without written permission of the author! All rights reserved.

All information contained in this book has been compiled to the best of our knowledge and has been carefully checked. However, this book is for educational purposes only and does not constitute a recommendation for action. In particular, no warranty or liability is given by the author and publisher for the use or non-use of any information in this book. Trademarks and other rights cited in this book remain the sole property of their respective authors or rights holders.

Thank you so much for choosing this book!

www.ingramcontent.com/pod-product-compliance
Lightning Source LLC
Chambersburg PA
CBHW050232230526
45470CB00005B/1912